Property Investment Decisions

A quantitative approach

Stephen E. Hargitay

the Built Environment, University of the West of England, Bristol, UK

Shi-Ming Yu

*School of Building and Estate Management
National University of Singapore*

E & FN SPON

An Imprint of Chapman & Hall

London · Glasgow · New York · Tokyo · Melbourne · Madras

Published by E & FN Spon, an imprint of Chapman & Hall,
2–6 Boundary Row, London SE1 8HN

Chapman & Hall, 2–6 Boundary Row, London SE1 8HN, UK

Blackie Academic & Professional, Wester Cleddens Road, Bishopbriggs, Glasgow G64 2NZ, UK

Van Nostrand Reinhold Inc., 115 5th Avenue, New York NY10003, USA

Chapman & Hall Japan, Thomson Publishing Japan, Hirakawacho Nemoto Building, 6F, 1–7–11 Hirakawa-cho, Chiyoda-ku, Tokyo 102, Japan

Chapman & Hall Australia, Thomas Nelson Australia, 102 Dodds Street, South Melbourne, Victoria 3205, Australia

Chapman & Hall India, R. Seshadri, 32 Second Main Road, CIT East, Madras 600 035, India

First edition 1993

© 1993 Stephen E. Hargitay and Shi-Ming Yu

Typeset in 10/12 pt Times by Pure Tech Corporation, India
Printed in Great Britain by Page Bros (Norwich) Ltd

ISBN 0 419 16780 3   0 442 31659 3 (USA)

A catalogue record for this book is available from the British Library

Library of Congress Cataloging-in-Publication data available

♾ Printed on permanent acid-free text paper, manufactured in accordance with the proposed ANSI/NISO Z 39.48–199X and ANXI Z 39.48–1984

This book is dedicated to our families

Contents

Acknowledgements

We wish to thank our colleagues at Bristol Polytechnic, the National University of Singapore, University of Reading and the College of Estate Management, Reading, for their help and encouragement throughout the production of this book.

Our thanks are also due to Brigitte Hague, of Norwich Union Real Estate Managers, for providing part of Chapter 17, and to Ann Colborne, Associate Head of Department, Bristol Polytechnic and member of the Investment Surveyor's Forum, also for providing us with much of Chapter 17.

The authors are grateful to the Investment Property Databank Limited for permission to include the details of some of their products in this book. The responsibility of the presentation, interpretation and use of the IPD material rests with the authors. Finally, our thanks and appreciation is due to Maureen Hargitay, for her patience in typing and editing the manuscript.

Preface

Investment in general, and property investment in particular, have been traditionally regarded more as an art than a science, where investors, decision-makers and analysts rely more on their experience, subjective judgement and 'feel' rather than objective, quantified evidence and sophisticated analytical procedures. Investments are made, and will continue to be made, on the basis of what investors believe, or expect, to happen in the future. Whilst subjectivity can never be eliminated from the investment decision-making process, a rational approach to the formulation of investment strategies and tactics is essential.

The importance of property has increased significantly amongst other investment media over the past thirty years. The amount of capital invested in property and diverted away from other investment opportunities is huge and requires efficient stewardship. The people responsible for planning further capital investments in property, and for the continuing maintenance and improvement of the investment efficiency of existing assets are the managers of property funds or portfolios and their respective expert advisers.

Most property fund managers have a conservative outlook and are adverse to change as far as their management and analytical methods are concerned. Tradition and fashion remain the principal determinants of their attitudes to decision-making concerning new investments and the management of existing assets. This traditional approach, however, simply cannot cope adequately with the complexities of the present-day investment scene. All those involved with decision-making, or with the provision of expert advice to investors in property, need to understand the fundamental relationships in the investment process. They need to be familiar with the theory and methodology currently used and accepted in the various alternative investment media, so that property investment decisions can be reconciled and put on a par with investment decisions made in alternative, and sometimes competing, investment media.

The aim of this book is to lay down the theoretical foundations of investment decision-making, incorporating the techniques and procedures of modern management science, so that particular decisions regarding property investments can be made efficiently and rationally.

Stephen Hargitay, Bristol
Yu Shi-Ming, Singapore

About the authors

Stephen Hargitay is a principal lecturer and Director of Studies in the School of Valuation and Estate Management at the University of the West of England, Bristol. He is also a visiting lecturer at the University of Reading and a tutor/examiner in Property Investment at the College of Estate Management, Reading. He has been involved with computer education since 1963, and the education of general practice surveyors since 1971. He has developed a number of computer applications for the property industry and was engaged by several organizations as a consultant. He has published a number of articles on property investment topics; he is the author of the following publications: *A Guide to the Hungarian Real Estate Market* (RICS Books, 1991), and with Tim Dixon and Owen Bevan, *Microcomputers in Property* (E. and F.N. Spon, 1991), and with Tim Dixon, *Software Selection for Surveyors* (Macmillan, 1991).

Yu Shi-Ming is a senior lecturer and Course Director of the BSc Estate Management Course at the National University of Singapore. He has been involved in a number of real estate consultancy projects in Singapore and in the People's Republic of China. He is a member of the Editorial Board of the *Journal of Property Investment and Valuation*; he has also edited two annual reviews of the property market in Singapore.

Introduction

The **investment decision** is concerned with the acquisition or disposal of investment assets. The assets may be **real assets** or **financial assets**. Real assets include land, buildings or interests in land and buildings, plant, machinery, stocks of material, etc., whilst financial assets are various forms of securities, deposits, debt instruments, etc. Most investors possess investment portfolios which are a mixture of financial and real assets and their interactions within the portfolio cannot be ignored.

The investor usually faces another set of problems, which is closely related to and emerging as the companion of the investment decision. These problems are to do with the decisions associated with the financing of the investment projects contemplated. The **financing decision** is expected to resolve the question of how much money should be raised, and in what ways, for the investment projects proposed. The **financial markets** provide the means through which the investor may have access to finance in various forms. The understanding of the **money** and **capital markets** therefore is extremely important to the investor if he is to resolve his funding problem in an optimal way.

The decision-making process requires the establishment of criteria against which investment projects and propositions could be evaluated; it also needs the existence of alternatives from which the selection is to be made or an order of preference be drawn up. The criteria and the alternatives will be perceived in **value terms**. In the world of investment, value is usually expressed in **money terms**. The fact is that the base of decision-making, relating to investment, is quantitative and therefore lends itself easily to rational treatment, although, in practice, the processes of investment decision-making are often irrational and riddled with inconsistencies.

The investment characteristics of property are significantly different from the characteristics of assets in other investment media. This is the reason why property is so useful and attractive for the purposes of diversification. On the other hand, such differing characteristics isolated property from the other media, in which tremendous strides were made in the development of decision-making methodology and the modernizing of investment and portfolio management techniques.

The full integration of property into the global investment portfolio depends on the full understanding of the investment characteristics of property not only in isolation, but also in the portfolio context.

A successful way to achieve the integration of property into the global investment portfolio is to collect all the property assets into a specialized property portfolio, then treat the property portfolio as a single asset in the global portfolio. The property portfolio is then assembled and managed by property experts who thoroughly understand the characteristics of the assets in their care. If the managers of property portfolios are also equipped with the understanding of the rationale of modern portfolio management practice, the efficiency of property portfolios as investment vehicles should be virtually guaranteed.

As the unitization and securitization of property assets gain ground, the complete integration of property into the global investment portfolio will occur, provided that the investment managers and property experts understand each other's problems and methodology.

The first step on the way to integration must be the recognition of the special investment characteristics of property, the understanding of the process of construction of property portfolios and the identification of the property portfolio problem.

The second step is the examination of the theory and methodology which have evolved in other investment media in order to see if that theory and methodology can be used or adapted for property investment and portfolio work.

Until quite recently, portfolio activity was regarded as an art, where intuition and 'feel' dominated decision-making. Objective, quantitative analysis was not possible as the theoretical foundations were not yet laid. Modern **portfolio theory** was pioneered by Harry M. Markowitz, who in his article 'Portfolio selection' laid down the theoretical foundations of the rational approach to the selection, analysis and management of investment portfolios. He postulated the concept of efficient diversification, and from his seminal work portfolio theory rapidly developed. Portfolio theory and the **capital market theory** now form a fairly coherent, theoretical framework ready for implementation in practice.

There are five sections in this book. Part One is devoted to the general aspects and setting of property investment decisions. Investment usually means the acquisition of assets by the investor with the view of satisfactory returns in the future. The capital committed to the acquisition of assets and the expected returns are exposed to risk. Generally the greater the exposure to risk, the higher will be the rate of return expected by the investor as a reward for bearing the risks involved. There is a variety of different types of asset from which the investor may choose in order to achieve his objectives. The main asset types available to the investor are cash assets, financial assets and real assets. The selection criterion of individual assets from these asset types is based on the provision of an adequate rate of return at an acceptable level of risk exposure.

Major investment decisions should be made on the basis of analysis which takes into account the state of the economy, the performance of an industrial sector, the financial and market status of the company, in the valuation of a company's share and the prediction of its expected future profits and dividends.

When the investor is active in the stock markets, his objectives may be achieved over short time periods. Profits and gains can be realized relatively quickly.

When the investor contemplates investment in property, it is likely that his aspirations regarding incomes, profits and gains will take a relatively long period to achieve.

Setting aside speculation, the value or worth of an asset is dependent on the net income stream derived from it. The analysis of the income stream and its appropriate conversion into capital values is the central core to investment valuations and appraisals.

The methodology of investment appraisal has evolved from various investment media where the evaluation of investment alternatives presented some special problems. The equity, gilt and property investment markets have all produced appraisal procedures and practices which worked reasonably well in the isolation of a particular media, but presented great difficulties when investment decisions were to be made reaching across the boundaries of the media.

When an investor buys an asset, he expects to receive future benefits in the form of income and cash receipt from the eventual, future sale of the asset. The investor expects the future benefits to be greater than his outlay. The purchase of the asset therefore is an investment decision made on the basis of the capital outlay and the estimate of future benefits or rewards.

Market Value is an estimate or forecast of the transaction price which would be most likely to occur under perceived market conditions.

Investment value is an estimate or forecast of what a rational investor would be prepared to pay for an asset under specified conditions and investment criteria.

The best investment vehicle to attain the general investment aim of maximizing returns whilst minimizing risks is a carefully selected and expertly managed investment portfolio. The selection and management of the portfolio is the most crucial part of the investment process. It is necessary therefore that the selection and management process is thoroughly understood and decision-making methods are adapted for these important purposes.

Part Two concentrates on various **quantitative decision techniques**. Decision analysis or the science of decision-making is concerned with the study of factors or information which affect the decision problem, as well as the evolvement of techniques which attempt to clarify and analyse the problem in such a way as to increase the chances of attaining consistent and acceptable results. To improve the quality of decision-making the three approaches are: first, better understanding of theory of decision-making under uncertainty; second, application of techniques and systems which will help to improve the process; and third, improve the quality and increase the quantity of information.

Quantitative techniques are useful aids for arriving at correct decisions and in demonstrating the basis of those decisions to others. The fundamental principle of the techniques is to allow the various relevant factors to be identified, quantified and then combined with the objective of achieving a rational analysis of the problem. Such techniques are also useful to the manager, in that they structure his thought processes and help to ensure that all the relevant factors in each case are considered.

There are also many situations in the real estate industry where scarce resources must be allocated amongst a number of demands which necessitate optimization. Resource allocation and planning methods seek to help decision-makers optimize their resources to give them the best returns. In addition, the management of resources is also important if decision-makers hope to maximize efficiency and therefore returns.

Whilst the real estate industry has been generally slow in taking up a wide range of quantitative techniques, one area that has been applied extensively is the network planning method.

Risk is a principal feature of all investment activities and investors throughout history have made attempts to eliminate it. However, investors soon realized that the total elimination of investment risks was virtually impossible and turned their attention to the reduction of risks. The idea of diversification amongst investments to reduce the overall investment risk is thousands of years old; although the original concept of diversification was intended to provide a safeguard against the total and catastrophic loss of the invested capital, it has evolved during this century into a sophisticated rationale for the management of investment risk.

Uncertainty and risk are, unfortunately, inescapable and a prevailing aspect of investment. Modern statistics and decision theory provide procedures for the making of investment decisions under uncertainty. The understanding of the basic concepts of probability is essential in the management of uncertainty. Through the application of the rules of probability and relatively simple statistics, the quantitative analysis of risk can be attempted. Decision trees are very useful in the analysis of complex investment decision problems.

Part Three concentrates on **investment performance measurement**. Performance measurement and analysis have attracted considerable attention, particularly in recent years. A whole industry has developed offering professional advice concerning the measurement of portfolio performance and the interpretation of the results of the measurement.

Most of the activity and development has so far been limited to the performance measurements of investment portfolios containing stock market securities. Although property assets represent a substantial portion of the global institutional investment portfolio, it has been considered too complex to include them in the overall performance measurement of funds. Property assets and property portfolios have been treated separately from the point of view of investment performance.

In Part Four the **information needs** and **information provision** and management for effective decision-making are examined. Investment decision-making, the appraisal of investment propositions and the measurement of investment performance require a considerable amount of information in a quantitative form. The interpretation of such quantitative information and its use requires the understanding of a number of mathematical and statistical procedures. The degree of complexity of these mathematical and statistical methods can be quite daunting for the investor. Fortunately, these complex mathematical structures are only necessary for the theoreticians who need to explain the complex cause-and-effect relationships in the modern investment markets. For most investors, only a basic mathematical and statistical toolkit will be required.

Property investment analysis is basically an activity involving the use of information (input) to produce a service (output). It is therefore fundamentally important that the analyst knows what information is required and where to obtain it.

The demand for reliable, relevant and easily digestible information has been steadily increasing since the investment markets regained their composure after the Second World War. Information concerning the stock markets has been freely available and a sophisticated information industry has evolved around these markets. The official information provision of the stock markets is complemented by an army of professional advisers and the daily press also carries relevant, up-to-date information on current events in the security markets.

The recent explosive advances in technology have also acted as a catalyst in the re-modelling of the traditional channels of investment information. Information Technology now provides the means to maintain huge pools of information efficiently, affords access to information bases in a form tailored to the specific needs of investors and offers facilities to process the information provided. Advances in the field of information provision are such that they are increasingly viewed with real concern as decision-makers are overwhelmed by the deluge of information.

Part Five consists of examples drawn from practice, showing how **property-related decision problems** may be approached and resolved.

Property investment decisions

PART ONE

The rationale of investment and investment decision-making | 1

1.1 INTRODUCTION

This chapter reviews the fundamentals of investment as a wealth-generating activity. The investment problem and its management will be considered, together with the principal players, the decision-makers in this important economic activity. After the classification of the investment decisions, the investment decision-making process will be considered in detail.

1.2 PRINCIPLES OF INVESTMENT

The fundamentals of investment must be preceded by an attempt at defining investment as an economic activity. **Investment** is to do with the creation, enlargement and protection of wealth.

Investment involves the commitment of a capital sum for benefits to be received in the future in the form of an income flow or capital gain or a combination of both. As an economic activity, investment may be defined as follows: *Investment is the utilization of capital resources for maximum possible reward.*

Funds are committed to various ventures which promise attractive returns at the price of risking a partial or total loss of the funds without an absolute guarantee of the size or receipt of returns. Investment therefore represents **certain sacrifices** for **uncertain benefits**.

Investment has two principal aspects: anticipated return and risk. The return aspect is perhaps the easier one to perceive and measure. It is usually translated into money terms and its size is all-important. The risk aspect, on the other hand, presents serious conceptual and analytical problems. Risk is a difficult concept to perceive and define and its direct measurement is virtually impossible. In an ideal, rational world the investor is expected to live by the maxim: *'Maximize returns whilst minimizing risk'.*

Unfortunately, in the real world investors have difficulty in adhering to this discipline. In the first instance, investors do not always behave rationally as they are usually influenced by the vagaries of the markets in which they are active and are prone to follow fashion. Furthermore, the problems associated with investment risk cannot be totally eliminated, even in an ideal world.

Investments can be classified in many different ways and there is a wide spectrum of investment opportunities available to investors, all with different investment characteristics. This diversity creates the problem of choice, requiring the establishment of criteria and a rational basis for comparison. A rational choice is usually made on the basis of return – risk characteristics, provided they are measurable against some suitable standard.

The primary problem for an investor is the assessment of desirability of the acquisition or disposal of an individual investment proposition. In addition, he may have to face further complications:

1. When the funds available for investment are limited – he has a *budget problem.*
2. When surplus funds are to be invested but suitable investment projects are not readily available – he has a *market supply problem.*
3. When funds committed to certain projects cannot be realized and switched to other projects easily – he has a *liquidity problem.*
4. When the return–risk attributes of investment propositions must be considered in relation to the rest of the assets owned – he has a *portfolio problem.*

1.3 THE INVESTMENT PROBLEM

The investment problem may be defined as a problem of decision-making in the presence of uncertainty and risk. The investment problem is a complex of the following three fundamental problems:

1. *Problem of selection* Choose the investment alternative which promises attractive returns at an acceptable level of risk. This implies the need of the quantification and measurement of the return and risk expectations.
2. *Problem of allocation* Decide the appropriate level of capital commitment, taking into account the degree of exposure to risk.
3. *Problem of timing* Decide the timing of the acquisition or disposal of investment projects, in attempting to achieve some return targets whilst minimizing the exposure to risk.

These problems are further complicated when the extra dimension of the interdependence of individual investments need to be considered in a portfolio setting.

1.4 TOWARDS THE MANAGEMENT OF THE INVESTMENT PROBLEM

At the core of the investment problem lies uncertainty which inevitably affects all future events. This problem of uncertainty can never be resolved completely as nobody will ever have a complete knowledge of the future. However, the problem of uncertainty can be managed as the consequences of the actions taken in the past, present and in the future are not completely random. Some consequences are not only foreseeable, but inevitable, barring some catastrophic event. On the other hand, the circumstances in which such consequences would be evaluated in the future are subject to random variations.

It is safe to assume that the primary aim of investors is to maximize the return on the capital committed. Whilst trying to safeguard their capital, in real terms, they wish to ensure the receipt and size of future returns, also in real terms. All rational investors would prefer higher returns to lower returns and lower risks to higher risks.

Throughout the ages, investors approached these problems in a number of different ways. Some of these approaches were wise and rational, others were quite irrational. The dictum 'Do not put all your eggs in one basket' appears amongst the proverbs of most languages. Some investors recognized the danger of the 'all or nothing' approach and saw the safety aspects of having a collection of investments. Other investors opted for another promising way to tackle the investment problem. Confidence in their experience, special skills and expertise has led them to the 'single basket' or 'specialist approach'. Their strategy was formulated by Andrew Carnegie, whose maxim was: 'Put all your eggs in *one* basket and then *watch* that basket.'

Only a few investors are in the position constantly to watch and manage their 'basket'. In most cases, the sound, rational strategy is to diversify, as risks may be reduced by a trade-off with return. This risk–return trade-off may be illustrated through the case of investment in government securities where the investor accepts a lower return for the virtual freedom from risk.

Large institutional investors usually diversify their investment portfolios on three levels:

1. They diversify amongst investment media.
2. They diversify amongst the sectors of a particular investment medium.
3. They diversify amongst individual investment projects within a particular sector or a particular investment medium.

The investment problem demands continuing attention from the investor and its successful management remains the most fundamental task of investment managers and their expert advisers.

1.5 TYPES OF INVESTMENT

In all civilized societies money is the life-blood of the economy. It appears to be the most important commodity and its supply and allocation is extremely important. Business and industry need an increasing supply of this particular commodity and their requirements over the long term must be met by investors and – to obtain the necessary injection of money – they must sell stocks and shares to the public. There is also a demand for money supplied by investors to enable governments and local authorities to carry out their various activities. Although governments and local authorities raise funds by taxation, nevertheless, without the investor's money, most of their schemes would encounter extreme difficulties.

The demand for investment to provide the flow of funds into various economic activities is usually satisfied by the individual investor and by large investing institutions. The following principal types of investments can be identified:

1. Money which is lent creating a debt. Such debt may be payable on demand, or at a specified date in the future.
2. Investment which transfers to the investor a part or share of a business enterprise.
3. Indirect investment in stocks and shares or in property through the purchase of an interest in a managed fund.
4. Investment in property interests.
5. Investing in non-yielding investments such as chattels, works of art or various commodities traded in commodity markets.

Most investments, particularly those made by institutional investors, would fall into the following categories:

1. *Fixed-interest stocks* representing marketable debt; government stocks, local authority stocks, company debentures and loans belong to this category.
2. *Shares* representing the acquisition of a part of a business enterprise; preference shares, ordinary shares and some of the means to acquire shares in

companies such as Convertibles and Warrants and Traded Options belong to this category.

3. *Property investments* cover direct ownership of commercial, industrial, residential, agricultural property interests. Investors can also take an indirect stake in property by acquiring the shares of property companies, property bonds and property unit trusts. More recent developments in the acquisition of property interests are to become available through unitization and securitization.

The various types of investments are traded in specialized investment markets. A particular market is referred to as an investment medium. Some of these markets are highly sophisticated and efficient, whilst others are fragmented and highly inefficient. Market efficiency is usually measured in terms of the availability of the information regarding trading, in that market, and also in terms of the efficiency of the price mechanism.

Investors, in their quest to achieve maximum return from their investments whilst minimizing the associated risks, usually seek an appropriate investment vehicle. Such a vehicle may be a single investment or a combination of investments.

Stock market securities have generally been viewed as the principal investment vehicle; traditionally attention had been focused on this medium. The vast majority of theoretical and analytical work is the result of the continuing interest of investors in this field of investment opportunities. Alternative investment media, including property, are left with the modified and adapted theory and methodology evolved in the stock market.

1.6 THE INVESTORS AND MARKET MAKERS

The investors are usually classified under the following headings:

1. *Personal sector*, including the individual investor, executor and trustee.
2. *Commercial* and *industrial companies*.
3. *Banks*.
4. *Institutions*, including pension funds, insurance companies, investment trusts and unit trusts. All investment media are dominated by the institutions. As far as the stock market is concerned, pension funds own nearly one-third of all UK quoted shares worth around £150 billion. Millions of individuals have a stake in these institutions and they provide the regular inflow of funds for the institutions to invest. All the nationalized industries, local authorities and private companies have pension funds who are major investors. Insurance companies account for another quarter of UK quoted shares. Investment trusts and unit trusts also have a powerful presence in the stock market. The institutions influence the stock market substantially.

In the personal sector, more and more private individuals seek investments, primarily in the field of personal savings, house purchase and increasingly in the stock market. Until 1984 there was a steady decline in the number of private shareholders. However, individual share ownership is rapidly gaining momentum through the growth of employee share schemes and the privatization of industries. Private shareholders now account for around 12% of the adult population in the UK.

The most sophisticated investment market is the Stock Exchange whose members deal for investors wishing to buy or sell stocks and shares. In this market the dealings were done between brokers and jobbers who were also regarded as market makers. Since the 27 October 1986 there is only one class of member of the Stock Exchange, called the broker/dealer. On the same date, some major changes occurred in the traditional ways of the stock market. It was brought into line with international practice of 24-hour trading around the globe and also allowed large international firms of share dealers to become members. Members of the Stock Exchange buy and sell shares continuously, under a strict code of conduct.

The brokers are the agents who trade in stocks and shares on behalf of investors. They undertake to buy and sell shares in listed companies, government stocks and other forms of securities at the best possible price to the investor, process the transactions, provide investment advice and offer investment management services.

The market makers are those members of the Stock Exchange who are wholesalers, competing with each other to supply brokers with stocks and shares and also buy shares from brokers. Since the 'Big Bang', the traditional distinction between brokers and market makers has virtually disappeared.

1.7 CLASSIFICATION OF INVESTMENT DECISIONS

Investment activity can be regarded as synonymous with decision-making, since investment is a commitment of resources in view of future returns, and decision-making is a commitment to an alternative activity which appears to be most likely to achieve a definite objective. Investment decisions are made by investors in an environment which contains a number of factors influencing the attitudes of decision-makers. Figure 1.1 shows the investment decision in its environment; the decisions made by investors may be regarded on two levels: the strategic level and the tactical level.

The **strategic decisions** relate to the setting of the overall policy objectives, the selection of investment media and the allocation of funds to selected media.

Tactical decisions relate to the implementation of strategic decisions within an investment medium. These decisions are about the selection of sectors of

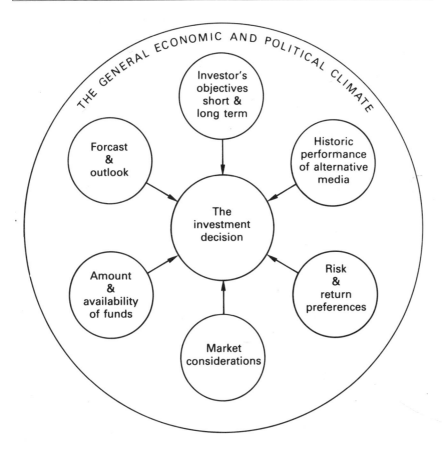

Fig. 1.1 The investment decision and its environment.

an investment medium and the choice of individual investment projects within the selected sectors.

Strategic and tactical decisions are subdivided into selection, allocation and timing decisions. The general classification of investment decisions is shown in Table 1.1.

Table 1.1 Classification of investment decisions

Decision	Strategic	Tactical
Selection	Selection of investment media	Selection of sector and individual securities
Allocation	Allocate funds amongst selected media	Allocate funds amongst sectors and securities
Timing	Switching funds between investment media	Acquisition and disposal of individual securities switch sectors

1.8 THE INVESTMENT DECISION-MAKING PROCESS

The **investment decision-making process** consists of the following five steps:

1. Definition of objectives and specific goals.
2. Search for a set of alternative investment projects which promise to achieve the objectives and goals set.
3. Evaluate, compare and rank the alternatives in terms of the quantified expectations of risk and return.
4. Choose the most satisfactory alternative.
5. At a later date, evaluate the consequences of the decisions taken earlier, draw conclusions, revise goals and criteria.

Although the above steps appear to be straightforward, there are a number of difficult questions to be resolved with each step.

It is appropriate to make a distinction between objectives and goals. The aims of the overall investment strategy are expressed in the objectives set by the investor. These statements of objectives do not set out to show specific goals, but they provide the overall framework and constraints for the lower, tactical levels of decision-making.

Goals, on the other hand, are specific statements about the desired consequences of investment decisions on the tactical level and they also influence the making of management and operational decisions.

Goals are formulated in terms of:

 (i) target or minimum acceptable rate of return;
 (ii) acceptable level of risk;
(iii) growth requirements with regard to income and capital value;
(iv) liquidity preferences.

The explicit statement of a goal requires the quantification of targets as single value items or as value ranges. Specific tactical goals must be quantified by a statement about the availability of funds for the investment and an appropriate time horizon within which the targets are expected to be achieved.

The search for investment alternatives, in certain investment media, may be seriously affected by the dearth of suitable proposition.

The evaluation and comparison of the alternative propositions present problems of the appraisal methodology. These problems are particularly acute in respect of the quantification of risk expectations. The choice of an alternative depends, amongst other things, on the use of the appropriate criteria. It is most important that the criteria are rational and theoretically sound, yet simple and pragmatic. The principal problem here is, once again, the presence of uncertainty which tends to reduce the effectiveness, even of the best criteria.

The continuous monitoring and periodic review of performance are essential if decision-making is to be effective in maintaining a course towards the investment targets set; Fig. 1.2 illustrates the investment decision process.

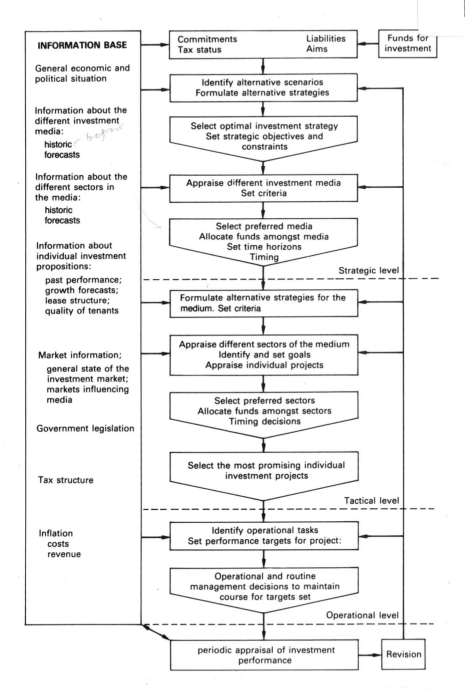

Fig. 1.2 The investment decision process.

The theoretical framework for rational decision-making is now in an advanced state. Since the end of the Second World War, decision theory has developed into a sophisticated discipline. As a result, decision-makers today have an amazing variety of theoretically sound, quantitative decision-making tools available to them. In practice, regrettably, only a small fraction of these tools have been used to date. Increasingly decision theory is being reluctantly accepted as it appears the only way to resolve complex investment decision-making problems in an increasingly complex and dynamic economic environment.

1.9 SUMMARY

Investment is the use of resources in a rational way in order to achieve maximum rewards in the future. Because of the incomplete knowledge about the future, the rewards from investment are subject to uncertainty. Investors risk their resources in order to achieve their objectives. The principal difference between investment and gambling is that whilst rational investors weigh up the risks involved and the rewards expected, gamblers will stake their resources on some chance and speculate recklessly.

If investment is assumed to be a rational activity, then decision-making associated with it must also be placed on a rational basis. The investment problem therefore should be viewed as a complex, rational decision-making problem, to be managed in the presence of uncertainty. The investment problem disaggregates into the problems of selection, allocation and timing.

The management of the investment problem is a conscientious and continuous effort towards the achievement of objectives and goals perceived, in terms of returns and risks.

It should be recognized that the practical decision-maker rarely has time for elegant theories; he wishes to make good decisions but in the real world he is faced with a shortage of reliable information, inadequate appraisal methodology and inefficient decision criteria. Under these circumstances, decision-makers facing the problems associated with investments could benefit from the use of simple but theoretically sound decision tools.

In the present investment scene, with the advent of Information Technology, investors are increasingly facing the problems associated with too much information being available to them, which will inevitably lead to confusion, loss of direction and inefficient decision-making. It is becoming increasingly important to safeguard investors from the deluge of sometimes totally irrelevant information and enable them to identify the essential information requirements for rational judgements in this increasingly complex, 'high-tech' investment environment. This is why (more than ever) the theory and methodology associated with investment activity must be understood by all who, at some point, will be called upon to exercise their judgements, set objectives and criteria in the commitment of resources for future gain.

FURTHER READING

Frost, A.J. and Hager, D.P. (1986) *A General Introduction to Institutional Investment*, Institute and Faculty of Actuaries, Heinemann, London.

Fraser, W.D. (1984) *Principles of Property Investment and Pricing*, Macmillan, London.

The Stock Exchange (1986) *An Introduction to the Stock Market*, London.

Sprecher, C.R. (1978) *Essentials of Investments*, Houghton Mifflin, New York.

2 Investment strategy and objectives – the portfolio approach

2.1 INTRODUCTION

Investment usually means the acquisition of assets by the investor with the view of satisfactory returns in the future. The capital committed to the acquisition of assets and the expected returns are exposed to risk. Generally the greater the exposure to risk, the higher will be the rate of return expected by the investor as a reward for bearing the risks involved. There is a variety of different types of asset from which the investor may choose in order to achieve his objectives. The main asset types available to the investor are cash assets, financial assets and real assets. The selection criterion of individual assets from these asset types is based on the provision of an adequate rate of return at an acceptable level of risk exposure.

Most rational investors are risk-averse – i.e. they prefer less risk to more risk and more return to less return. The rationale for the management of investment risk has been established long ago as some form of diversification amongst various investment assets. As the writer of the Babylonian *Talmud* put it succinctly some 1500 years ago, 'Man should always divide his wealth into three parts: one-third in land, one-third in commerce and one-third retained in his own hands'. Thus the **portfolio approach** to investment can be traced back into ancient history. The portfolio is a combination of several investments assembled for the purpose of the management of investment risk and for the enhancement of investment returns.

Investment decisions are usually made on the basis of fundamental analysis which takes into account the state of the economy, the performance of an industrial sector, the financial and market status of the company, in the valuation of a company's share, and the prediction of its expected future profits and dividends. Technical analysis, which involves the charting and analysis of historical price movements to predict future price movements, is often used in practice to underpin the conclusions of fundamental analysis.

Modern portfolio theory (MPT) provides another more rational method to assemble a portfolio of risky securities. MPT provides the rationale to select a combination of risk-free assets and risky assets that would meet the investor's goals and objectives. The first and fundamental problem facing the investor is to establish his investment goals and objectives. Only after the clear definition of the investment goals and objectives can the appropriate investment strategies and tactics be worked out.

This chapter concentrates on the **portfolio problem** and its components from which the investment goals and objectives may be articulated and suitable portfolio selection policies and portfolio management strategies formulated.

2.2 THE PORTFOLIO PROBLEM

The **portfolio problem** is defined as: *a problem of choosing a collection of individual investments, or bundles of investments, that taken together have the most desirable characteristics with respect to risk and expected return.* Since the components of a selected portfolio have individual and transient performance characteristics, the portfolio problem extends beyond initial selection into continuing analysis and revision – i.e. management. Although selection and management are two distinct problem areas, they are mutually interdependent. The prime objective remains the same throughout both stages: optimize the trade-off between returns and risks. This objective is pursued through diversification which is expected to lead to a reduction of the overall investment risk and also, in certain circumstances, could lead to an increase in the value of the portfolio.

As the investment performance of the individual components of the portfolio vary continuously and, at the same time, the environment in which such performance is evaluated and goals re-assessed is also subject to change, the portfolio problem does not have a clearly defined solution. The portfolio problem, however, can be managed towards the achievement of some perceived investment targets and objectives. Such targets and objectives are *optimal*, and only exceptionally optimum solutions. For these reasons, it is more appropriate to pursue the effective management rather than the solution of the portfolio problem. If the management of the portfolio problem is to be effective it should have:

(i) a sound rational basis for the selection of an efficient set of investments which will satisfy the investor's objectives within his overall investment strategy at a particular point in time;
(ii) a sound basis for the revision of the portfolio by using rational methods for the measurement and analysis of past portfolio performance and for the assessment of likely future performance.

The effective management of the portfolio problem hinges on the establishment of investment goals and objectives and the planning of strategies and tactics to reach the objectives set.

In order to establish realistic investment goals and objectives, the investor's needs must be clearly identified, together with his attitude to risk bearing. The investor must have funds to invest, and he must be in a sound financial position to accept some risks associated with the investment activity. The effective management of the portfolio problem is virtually impossible without adequate knowledge of the risk and return characteristics of the various securities and assets that are available for acquisition, or already included in the portfolio.

2.3 THE ESTABLISHMENT OF OBJECTIVES

The first step in the establishment of objectives is the determination of the **investor's needs**. The needs of an individual investor will be substantially different from those of an institutional or the corporate investor. Nevertheless, the primary motive for investment in all cases is profit.

The individual investor's needs are partly financial and partly psychological. Usually individuals invest to provide for retirement, educating their children or simply hoping to increase their wealth through investing. The psychological approach of individuals to money matters, particularly bearing risks, varies greatly. Some have the attitude that since they only hazard a small amount of money when investing in risky ventures, the risk aspects of investment are insignificant, so they accept risks quite readily. These individuals are psychologically suited to investment in the stock market. Other individuals would regard the loss of their small investment as a major disaster and even the normal fluctuations of the market prices of stock market securities causes them great distress. These individuals are not suited to risky stock market investments and should keep their monies in National Savings Certificates or building society accounts.

The institutional investor usually invests to make adequate provisions to cover his future liabilities. These liabilities must be met as and when they arise. Since it is virtually impossible to produce returns which would be equal to and coincident with liabilities, the institutional investor needs returns which accumulate a surplus over the expected future liabilities. Various institutional investors have distinctly different liabilities and tax status and therefore require different investment and portfolio structures to meet their respective commitments in the present and in the future.

The corporate investor's needs are different again. The corporate investor is particularly concerned about the efficient management of current assets, particularly cash. A healthy cash flow is a prerequisite of success in business. As the opportunity cost of holding cash rises with the rising of the rates of return on securities, it is prudent to invest any cash surplus in marketable securities. Corporate investors also plan ahead to meet the future costs of the changes in product ranges, new plants, competitions and takeover battles. For these needs, the existence of a portfolio of marketable securities would be very useful indeed.

The formulation of the portfolio objective involves the listing and ranking of the principal needs and constraints for a particular investor. The list of the principal needs and constraints are as follows:

Needs
   (i) Need for the security of the capital invested. This can be expressed in current money terms if the portfolio may have to be liquidated at a short notice, or in constant money terms by taking inflation into account.

  (ii) Need for the security and stability of income. Again, this can be expressed either in current money terms or constant money terms by making the appropriate adjustments for inflation.

 (iii) Need for the readily marketable investments, i.e. liquidity.

 (iv) Need for tax exemptions or reliefs.

  (v) Need for external professional management.

Constraints
   (i) Limit on the size of immediate capital commitment.

  (ii) Limit on the size of future investments. These future capital commitments are either regular or random.

 (iii) Upper limit of the level of risk bearing.

 (iv) Minimum acceptable rate of return.

  (v) Term of investment.

 (vi) Statutory controls.

It is extremely important that the investor sees his particular position amongst these needs and constraints and is able to express their relative importance. Only then will he be in the position to articulate his portfolio objectives clearly.

2.3.1 Security of capital

In most cases, this aspect should be the most important consideration, as the preservation of the value of the portfolio is the prerequisite of future income, and if the capital is lost, its replacement can be extremely difficult. A total loss of the capital is usually the result of recklessness or an unavoidable catastrophe.

Security of capital means more than just maintaining the original investment, it also means the protection of the purchasing power of capital. A partial loss of the capital, through the reduction of the real value of the portfolio due to inflationary effects, is a quite frequent phenomenon. Portfolios containing a significant proportion of gilts and Treasury stock are vulnerable to long-term interest rate fluctuations and the inflationary decline of money values. It is often suggested that common stocks tend to increase in value as the money values are deteriorating and therefore can be regarded as a reasonable hedge against inflation. However, common stocks are more risky in all other respects than fixed income securities; therefore the investor must try to strike a balance between the dangers of inflationary effects and other risks.

2.3.2 Security of income

Income is derived from the investment portfolio in the form of dividends and interest earned. The dividend income tends to be more erratic than interest income. For planning purposes the more stable income stream is preferred. The incomes derived are either used for consumption or re-invested. In either case, the planning of investment strategies and the articulation of objectives is much easier with stable income streams.

When the income is mainly used for consumption, the importance of maintaining the real value of the incomes becomes an important objective. When considering the long-term income needs, it is important to produce a buffer against inflation. This usually means an increase in the initial capital commitment, which is the only reasonably certain way to ensure the production of the required income in the long-term. Obviously there will be a surplus of income initially, but this surplus can and should be re-invested providing protection for the real value of the capital invested.

The main objective of the portfolio is either the provision of satisfactory income or the achievement of capital gains in the future. Accordingly, there is a distinction to be made between an **income portfolio** and a **growth portfolio**. The strategies appropriate for these differing objectives are significantly different and therefore it is important to define clearly whether the objective is income or growth.

2.3.3 Marketability and liquidity

Marketability refers to whether a particular asset can be bought or sold easily, whilst liquidity usually means the ease at which a particular asset can be turned into cash. Both these aspects are important to those investors whose objective is to take advantage of attractive opportunities that may arise. Most investors follow the rule of buying stocks when the prices are low and selling when prices are high. Purchasing stock at the opportune moment will require the quick conversion of some of the existing assets into cash or cash equivalent.

Marketability depends, to a certain extent, on the place where assets are traded. Assets listed on stock exchanges are generally more marketable than others traded locally or on the OTC (over- the-counter) markets. Stocks of large companies are more easily marketed than those of the smaller ones, and the larger number of shares provide a more continuous market.

Liquidity is affected in a similar manner. The cost of conversion and transaction can be a considerable burden to the small investor. Large institutional and corporate investors usually transact large amounts of stocks and shares and their transaction costs are relatively insignificant.

Some portfolio managers set aside a portion of the portfolio to maintain a required level of marketability and liquidity, others use the surplus income from dividends and interest to purchase new assets when the opportunities arise.

2.3.4 Taxation

A very important factor in the selection and management of investment portfolios is the tax position of the investor. Most investment decisions can be made by considering the expected returns in the light of the investors' tax liabilities.

Investment returns are liable to income tax, corporation tax and capital gains tax. The tax status of individuals, corporations and institutional investors varies considerably, and the details of tax implications for various investors are outside the scope of this chapter. However, attempts to reduce the burden and impact of taxation have considerable implications on the setting out of portfolio objectives and the planning of investment strategies. It is sufficient to mention here that those investors who are liable to high rates of income tax must consider very carefully the various options available to them to reduce or defer the tax burden on their incomes. These investors should seek those investment vehicles which offer some tax relief or taxes already deducted at source. Alternatively, they may defer the tax burden by opting for returns in the form of capital gains. As a capital gain is made when the investment is disposed, the capital gains tax payment will be due after disposal. Future capital gains taxes can be reduced by utilizing the annual exemption allowance.

2.3.5 Professional investment management

Regarding the management of his portfolio and the burdens of investment decision-making the investor has a number of options. He can retain full control of his investments and make all the selection and allocation decisions himself. This option may be the cheapest in terms of fees payable to professional investment managers but it requires a considerable amount of time and knowledge to make it really effective.

The investor could retain some of the management functions by delegating the responsibilities of the day-to-day acquisition and disposal decisions to professionals such as the managers of unit trusts, etc.

The investor may decide to engage professionals to manage all his assets. There is a whole industry offering such services to investors. The investor should check whether his chosen manager is a member of the Financial Intermediaries Managers and Brokers Regulatory Association (FIMBRA).

Large corporate and institutional investors usually have specialist in-house investment departments. Nevertheless, they do use the services of external experts to enhance the efficiency and reliability of their in-house team.

2.3.6 Limits on the investment budget

Even the largest of institutional investors must consider the size of their investment portfolio and the volume of future funds flowing into it. Funds available for investment are determined by the total funds in the investor's control and the portion of the total funds needed for immediate business operations and for the meeting of current and future liabilities.

For an individual investor, the size of the investment budget is determined by the amount he can afford to invest from his annual income, taking his annual expected expenditure into account.

The investment budget of the corporate investor is determined by the size of the surplus funds over and above the operational costs and expenses.

Institutional investors, such as insurance companies and pension funds, have a continuing problem of finding suitable and satisfactory investment opportunities for a regular and large surplus of the inflow of insurance premia and pension contributions. The budgetary problems of these investors are characterized by the attempts to limit the exposure of the investor to a particular type of risk associated with certain industries or types of investment assets. For example, it is generally accepted that it is prudent not to expose the fund to the special risks associated with property by investing no more than 20% of the total value of the fund in property.

2.3.7 Limit of risk bearing

Most rational investors aim to minimize investment risks. The attitudes of investors to risk bearing is entirely subjective and very difficult to express in quantitative terms. The main problem with risk is that it is difficult to define and even more difficult to measure. Generally risk is regarded as the probability of not achieving a predicted rate of return.

The quantification of risk is usually achieved through the use of the (relatively simple) statistical measures of variability such as the variance and standard deviation of expected returns. Variability relative to the market is usually expressed with the beta coefficient. Quantitative measures of risk are essential for analytical purposes and for the setting of the limits of risk bearing. Unfortunately, these quantified expressions of risk cannot cover the whole risk spectrum and they are imperfect, hence most practising investors tend to treat them with scepticism.

Investors who are not particularly averse to risk tend to own equity portfolios with a large proportion of speculative stocks. More cautious and risk-averse investors usually aim for balanced portfolios containing fixed-interest securities and the equity portion of their portfolio is well diversified. A strongly risk-averse investor usually opts for a portfolio with gilts, convertible debentures and only a small, if any, equity portion.

Usually the smaller the size of the capital available for investment, the less scope there is for risk bearing. When the capital available is small, diversification is difficult and therefore the minimization of portfolio risk is virtually impossible.

If the investor is confident that he is well-informed about the variability of the returns on the securities already in his portfolio, or of those about to be acquired for his portfolio, the setting of the limit of risk bearing in terms of variability measures can be a relatively simple task.

2.3.8 Minimum acceptable rate of return

This constraint is closely tied to the investor's view of the security of income and capital in real value terms. Generally investors require higher returns from those investments with the higher risk content. In certain cases, the minimum acceptable rate of return is called the **threshold return** below which the investor's objectives and his ability to meet his liabilities would be in jeopardy. In the case of a pension fund, it is essential to achieve a certain rate of return on the invested funds, in the long run, if the future benefits of the contributors are to be guaranteed.

Investing institutions are in competition with one another for the inflow of monies from savers and small investors. Under these competitive conditions, it is important for them to be seen by the investing public as good performers. They therefore direct their investment managers to achieve a certain position in the **league tables**: position in league tables is dependent on their rates of returns achieved *relative* to others in the league.

Traditionally, return and yield expectations were based on the yield on gilts, all other forms of investments were expected to produce an additional risk premium, reflecting their relative riskiness. The rule was:

If	gilts yield	$x\%$
then the expected	yield on debentures	$x + 0.5\%$
and	yield on preference shares	$x + 1\%$
and	yield on ordinary shares	$x + 2\%$
and	freehold property	$x + 2\%$

During the period 1950–80 rising prices and interest rates considerably reduced the security of capital invested in gilts because they offered only a fixed return and both income and capital were seriously eroded in real terms. Consequently, the above-mentioned yield structure was reversed and ordinary shares offered lower yields than gilts.

Gilt-edged stock is regarded still as virtually risk-free and the more recent structure of yield expectations also is based still on the current yield on gilts; thus:

If	gilts yield	$y\%$
then the expected	yield on debentures	$y + (1 \text{ to } 2)\%$
and	yield on preference shares	$y + (4 \text{ to } 5)\%$
and	yield on ordinary shares	$y + (3 \text{ to } 5)\%$
and	freehold property	$y + (1.5 \text{ to } 2.5)\%$

These figures are, of course, only a rough guide; nevertheless, they are the subject of continuous debate.

2.3.9 Term of investment

The investor is likely to want access to some or all of the capital invested, at some point in time. As far as the individual investor is concerned, the length of time whilst his capital is tied up in investments is an important consideration. As far as large corporate and institutional investors are concerned, they need to determine the dates at which cash flow will be required. It is unlikely that these investors will be needing all their invested funds, nevertheless some of the foreseeable liabilities may be considerable. For example, if an insurance company is selling endowment policies which will mature in 10 and 20 years time, they would need to consider the purchase of securities which would match the expected liabilities.

2.3.10 Statutory controls

Most of the institutional investors are subject to some form of legal constraints regarding the way they carry out their business, the kind of securities they are allowed to invest in and the maximum volume of funds permitted in a particular type of investment.

The banks are regulated by the Banking Act (1979), which is intended to ensure sufficient liquidity and requires the banks to manage their assets prudently to achieve the most profits for their shareholders.

Building societies are controlled by the Building Societies Act (1962), which places them under the supervision of the Chief Registrar of Friendly Societies. The Chief Registrar controls the reserve ratio of the building societies, which expresses the general reserves as a percentage of total assets. The average reserve ratio of building societies is around 4% of the total assets.

Unit trusts must be authorized by the Department of Trade and Industry (DTI) under the Prevention of Fraud (Investments) Act (1958). Unit trusts are constrained by the DTI in a number of ways: investment in any one company is limited to 10% of any one class of capital and limits the value of the holdings in any one company to no more than 5% of the value of the trust fund. They are not allowed to invest more than 5% of the unit trust fund in unlisted securities. The 1958 Act also prohibits unauthorized unit trusts from distributing without special permission; property unit trusts belong to this particular trust category.

Investment trust companies (ITCs) are controlled with the Companies Act (1985). The prerequisite of the Stock Exchange listing of ITCs is that their investment in any one company should not exceed 10% of all of its investments and they cannot invest more than 15% in unlisted securities.

Insurance companies are supervised by the DTI and controlled with the Life Assurance Companies Act (1870), the Policy Holder's Protection Act (1975) and the Insurance Companies Act (1982). The aim of the legislation is to protect policy-holders from the insolvency of the companies. The DTI's control is complex and there are limits imposed on the proportions of assets allowed in any one security.

Pension Funds are normally set up as trust funds under a trust deed and they must invest their monies according to the Trustee Investment Act (1962). The first schedule of this Act is particularly important. Pension funds are also under the control of the Inland Revenue, in as much as they are regarded as gross funds – i.e. they do not pay income tax or capital gains tax.

2.4 PORTFOLIO STRATEGIES

The investment objectives are achieved through the appropriate policies and strategies. As far as the investment portfolio strategy is concerned, it divides into three well-defined areas:

1. *Selection*, which involves the selection of a particular type of portfolio which appears to be most appropriate to achieve the investor's objectives and the selection of individual assets for a particular portfolio.
2. *Allocation*, to decide the appropriate level of capital commitment to the portfolio as a whole and to sectors and individual assets in the portfolio.
3. *Timing*, for acquisition, disposal and restructuring of the portfolio and its components.

In the portfolio setting, these areas present the investor with an additional dimension of difficulty of the interdependence of the individual investment assets.

The principal objective of portfolio investment, to maximize returns whilst minimizing risk, is achieved through an appropriate strategy of **diversifica-**

tion. In order to put into effect a worthwhile diversification strategy, the investor needs to be in an appropriately strong financial position which is usually the privilege of large corporate and institutional investors.

There are two approaches to the planning of portfolio strategies. One is usually referred to as the **traditional approach**, which simply looks at the investor's objectives in terms of the need for income or capital appreciation and then selects those securities which appear to be the most appropriate to meet these needs. The other approach is a more theoretical one, and it is aimed to create a strategy which will maximize the expected returns on a portfolio for a particular specified level of risk. This approach is based on **modern portfolio theory** (MPT). The traditional approach of strategic planning is well respected in spite of all its idiosyncrasies and faults whilst the approach based on MPT is still treated with suspicion, mainly because of its heavy reliance on complex mathematics.

Traditional portfolio strategies are directed towards either the production of current income or to the achievement of capital gains in the future.

The **income portfolio strategy** emphasizes the need for the maximization of current incomes. Assets and securities selected for this kind of portfolio must be able to provide the income needs of the investor in current and real terms. The most sought-after asset is a common stock with high current income and with low risk profile. The securities that could provide the necessary incomes are bonds, preferred stocks and for the high tax-paying investor those securities which promise some sort of tax shelter. The portfolio will have a number of defensive securities, such as bonds and aggressive securities (i.e. common stocks and the balance between the aggressive and defensive components depends on the investor's view of the current state of the markets and the general economic climate, now and in the foreseeable future). The risk aspects of the income portfolio will include all the usual components of the total investment risks and investors will need to reduce these risks as much as possible.

The usual approach to the treatment of the total risk is to identify the different components of the risk and try to select those securities which promise minimum risk in a particular category – e.g. to guard against business risk invest in high-quality securities and government bonds, and to guard against the effects of inflation, invest in those common stocks which promise both income and growth. Some investors invest in securities which mature within a short period of time in order to avoid the risk associated with the prospect of rising interest rates. As the securities are redeemed, the money can be re-invested in the appropriately higher-yielding securities.

The **growth portfolio strategy** is mainly concerned with the increase of the future value of the portfolio. The principle is to place a greater emphasis on capital gains than on current income. Investors can achieve growth by re-investing the current income to build up the capital. Usually for this strategy those securities must be selected which have a high return and low-risk profile relative to the market. The treatment of the risk aspect can be a difficult one in this case as too many securities could fragment the total return, thereby the return provision will be insufficient. If the investor concentrates on only a few securities, the total risk may be unacceptably high. In this situation, the investor must be careful with the timing of the acquisitions and disposals to exploit opportunities when the prices are low and prepare to trade when the prices are high. Another approach to the minimization of risk is to attempt a continuing balancing act between risk-free securities such as gilts and Treasury bills and equities with a higher risk profile matched by greater growth potential.

It is essential, that the investor who follows a strategy pursuing a growth portfolio analyses the market and interest rates before setting out and monitors the behaviour of the market and interest rates during the build-up of his portfolio. Without the continuing monitoring, it is unlikely that this kind of strategy would succeed in the long term.

2.5 TIMING OF INVESTMENT ACTIVITY

The success in investment is dependent on the correct timing of investment activities. The investment markets exhibit cyclical tendencies in the long run, and they fluctuate according to the prevailing economic and political environment. The time-honoured rule of investors is to purchase stock at low prices and sell stock when prices are high or the stocks are considered to be overpriced.

Unfortunately, this is not an easy thing to achieve as it is impossible to predict with absolute certainty the state of the market. A number of investors follow a strategy of **buy-and-hold**: their argument is that, in the long term, the price paid for a particular stock will not matter as if the company is good enough the price of its stock will rise anyway. There is no need to rush into purchasing a particular stock at a particular point in time, funds can be put into bonds or fixed-income securities until the purchase of the stock is fully considered.

The timing of the purchase of securities can be resolved by setting target prices for each stock sought and buy when the stock is available at the target price. This purchasing strategy requires patience and it also presents some problems. The price of the stock may never go down to the target price or during the waiting period significant changes may occur in the economy or in the position of the company. There is always the possibility that changes may occur in the portfolio itself which could reduce the desirability of the stock in question. The problem of timing of the purchase of property assets is even more acute as desirable properties are not continuously available. The success of the target price based buying strategy depends on the continuous analysis of the portfolio, companies, market and the economy.

The strategy for the disposal of assets is equally important. Again, the correct timing of disposals is dependent on the analysis of the portfolio, economic trends and the recognition of the market cycle. The general rule is to sell securities when their prices are high. From the analysis of a security the present value of the future earnings can be assessed, and if the present value is much smaller than the current price, then the security should be sold. The analysis of a particular asset will also show whether or not the asset meets the original objectives. If the asset no longer contributes positively to the achievement of the portfolio objectives, the time is right for its disposal. The following points should be considered for the timing of disposals:

(i) asset no longer meets portfolio objectives;
(ii) market price of the asset is very high;
(iii) present value of expected earnings is considerably lower than current market price;
(iv) return and reward/risk ratio are low;
(v) other investment opportunities are extremely attractive;
(vi) tax advantages are to be realized.

Many investors find the continuing analysis of the markets and their own portfolio a great burden and prefer to use other 'recipes' to resolve the problems of investment purchases. These mechanistic rules are usually aimed to reduce the average purchase price to the lowest levels. The most well known of these mechanical investment plans are as follows:

(i) *Pound averaging* The investor selects a stock or a group of stocks and invests an equal amount of pounds (or dollars) at equal time intervals.
(ii) *Ratio plans* The investor reserves a certain proportion of his portfolio to fixed-interest stocks, to provide security and income, whilst the rest of the portfolio is invested in equities and other assets, to provide hedge against inflation and capital appreciation. The ratios are reviewed periodically. The investor buys and sells equities to maintain the pre-set ratios.
(iii) *Indexing* Stock is bought or sold if there is a significant change in the appropriate stock market index.

The pre-requisite of these investment plans is that the investor adheres to the rules once he accepts them. A further advantage of these mechanistic investment plans is that they can be encoded in computer programs, which reduce the computational chores to a minimum and make decision-making virtually automatic.

2.6 SUMMARY

The essential ingredient for success in investment is planning. The setting of realistic objectives and the recognition of the constraints must precede the acquisition of assets and the assembly of the investment portfolio. The needs of the investor must be recognized before realistic targets and objectives can be set. The ranking of the objectives is also extremely important.

Most institutional and corporate investors are constrained in their investment activities. The statutory constraints and regulations are very important indeed and must be incorporated into the portfolio planning process.

The clear statement of objectives and constraints is essential for the formulation of appropriate investment policies and strategies. The portfolio strategy covers three aspects: selection, allocation and timing. All these aspects can be approached through the traditional ways, as described above, or the modern, more rational, analytical ways. Modern portfolio theory provides the rational albeit heavily mathematical approach of the future. The acceptance of these new ways by the investment fraternity is largely dependent on the education of the new generation of investors.

FURTHER READING

Firth, M. (1975) *Investment Analysis*, Harper and Row, New York and London.
Frost, A.J. and Hager, D.P. (1986) *A General Introduction to Institutional Investment*, Heinemann, London.
Kerridge, D.S. (1987) *Investment – a Practical Approach*, Pitman, London.
Sprecher, C.R. (1978) *Essentials of Investments*, Houghton Mifflin, New York.

3 | Decision criteria – return and risk

3.1 INTRODUCTION

Investment is defined as an activity which requires cash outlay with the aim of receiving, in return, future cash inflows. It is assumed that the increase in wealth, as a result of the purchase of financial assets, is the main objective of investment. The return on investment is a measure of the change in wealth resulting from the purchase of these assets.

Uncertainty is an important feature of investment. All investments involve the future which is unknown. This fact leads to one fundamental problem of investment: how can investment decisions be made under conditions of uncertainty? This question must be addressed by anyone who makes investment decisions. Uncertainty is in the nature of things, and those who enter into ventures which depend upon future outcomes must come to terms with it.

In spite of its importance, the effects of uncertainty are often neglected, or pushed aside as insoluble by investors. Failure to consider uncertainty and risk can have catastrophic consequences as investors could not only receive inferior returns, but also lose their invested capital.

Uncertainty and risk are usually considered as one and the same. An investment is considered risky because the investor is unsure about the actual return which he will realize from his investment, so that risk is related to the uncertainty of the future returns from an investment.

Not surprisingly, attempts have been made to develop a theory of decision-making which copes with uncertainty. The development of decision theory, particularly directed to solve the problems of investments in the stock market, is now a well-established and sophisticated discipline. The demand for the development of such theory has come mainly from financial institutions, particularly pension funds and insurance companies who require explanations and justifications for their various investment activities. The need for a rational approach to the treatment of uncertainty came with the 1974/75 collapse of the markets in the UK which exposed the serious shortcomings of the crude, 'back of the envelope' type approach. A more serious approach to the proper treatment of uncertainty and risk has also been made easier through the increasing amount of relevant information available and by the availability of relatively cheap and plentiful computer power for their proper analysis.

Investors need to identify the various elements of uncertainty, some of which may be within their control, whilst others lie outside their control. They also have to identify their attitude to risk. Some investors are more inclined to accept risk than others. Most rational investors would not be happy in taking risks without a careful identification and assessment of the risks involved. In investment practice, these risk assessments still tend to be very crude as the more sophisticated theoretical approach is still considered to be far too esoteric for most investors. The financial institutions however, who play the role of trustees to the investments of a large number of individuals, require a greater sophistication of the assessment of risks (Wilson Committee Report, 1980).

Good investment decisions, which can take into account the effects of uncertainty and risk, can only be attempted if and when the nature of uncertainty and risk is properly defined and suitable methods of quantification and measurement are established. In the subsequent sections of this chapter, investment risk and its components will be defined and methods of quantification and measurement of risk outlined. The traditional approach to the treatment of uncertainty and risk will then be contrasted with a proper and rigorous analysis and treatment of risk.

It is accepted that the principal purpose of business activity in general, and investment in particular, is to generate profits, or returns. The prerequisite of any analysis is to determine those factors which determine the level of profitability. Usually business, managerial and investment decisions are the first determinants of the level of profitability. The decisions are usually related to the following:

(i) The size of the operation with respect to the gross earnings and revenue produced.
(ii) The proportion of gross earnings and revenue absorbed by various costs.
(iii) The volume of funds required to support a certain level of business activity through investment in fixed assets and working capital.

The methodology of the measurement of earnings, profits and value changes must be clearly understood, otherwise the correct interpretation of the various measures and indicators will be virtually impossible.

3.2 THE MEANING OF RETURN

The return is seen as a reward for committing and exposing to risk some capital in investment projects. The total return has two principal components: the income earned, and the increase in the value of the asset acquired.

The part of the return which comes from the increase in the value of the asset is either **realized** or **unrealized**. This particular part of the return will be realized only at the disposal of that particular asset. This aspect could lead to some confusion when the investment performance of the asset is being analysed and monitored.

Investment returns are usually regarded as gross potential returns, although most prudent investors would be more interested in the **real returns** after allowances for the erosion of money values, taxes, etc.

As investments mean the purchase of future earnings, it will be necessary to formulate future return expectations. **Expected returns** are formulated from the examination of historical returns and the consideration of trends. The ranking of investment alternatives is usually carried out on the basis of the expected returns on the various alternatives.

In order to identify the total return required by investors, an appropriate yardstick or datum will be needed. This yardstick is usually a close alternative investment. The **target return** has two aspects: the return expected from a close, substitute investment, and the special characteristics, such as risk, liquidity, marketability, tax, etc., of the investment in question.

The return produced and expected in the future is the principal measure of the desirability of investment. The ability to produce net returns after costs and various other deductions is the meaning of **profitability**. The question of profitability is central to investment decision-making. The assessment of future profitability depends on the estimated current earnings and the trends and growth of earnings in the recent past. There are some conceptual problems in the measurement of earnings, particularly corporate earnings.

Most of the relevant information regarding current earnings and those in the recent past can be extracted from accounting reports, such as balance sheets and income statements. These accounting reports are prepared by accountants who strictly adhere to accounting principles and procedures. Difficulties arise when the returns and profits extracted from these reports need to be interpreted and subsequently used for decision-making. The problem is that accounting profits differ from economic profits. The reason for these differences is that the accountant will measure historic earnings, whilst the economist or investment analyst would be mainly concerned with the future flow of earnings over the expected life of a particular investment. The accounting periods which are essential for the preparation of traditional balance sheets are irrelevant to the analysis of the future earning profiles. Economic analysis is based on the concept of cash flow, whilst historic accounts are usually given on an accrual basis.

Opportunity costs are not deducted from earnings when accounting profits are calculated. The accountants usually treat interest costs as outgoings, whilst economic analysis requires interest costs, including opportunity costs, to be calculated on the equity capital employed in a particular investment activity.

Another danger-point is the treatment of the part of the return resulting from capital gains or losses. The accountant regards realized capital gains or losses as a component of income. The unrealized capital gains or losses are left out from the calculations.

The investment analyst, on the other hand, treats realized and unrealized capital gains or losses as part of the total return. Because of these differences between the procedures of the accountant and the investment analyst, great care must be taken that past earnings and returns are free from distortions and trends are properly recognized and interpreted.

3.3 THE MEASUREMENT OF EARNINGS AND PROFITABILITY

The objective of investment analysis is to enable the investor to select those investments which represent the best value for money. It is necessary to assess the value of the investments on offer, compare the assessed values with market prices in order to make the right decisions regarding acqusitions and disposals. The value of an investment is determined by its expected return and risk. Value assessments usually rely on subjective judgements based on measures of past performance. These measures reflect historic return profiles, current strength or weakness and future potential of various investment propositions. If comparative judgements are to be made, it is essential that the methods of measurement and the computational procedures are standardized as far as possible.

There is a wide variety of measures of earnings and profitability in use, particularly in the analysis of stock market securities. Some of these measures and indicators can be used albeit with due care in most fields of investment. Others are too specialized to be of use outside a particular investment market. There is now an increasing pressure to standardize the measures of earnings and profitability to cover most fields of investment activity in order to facilitate the proper comparison of opportunities in competing investment media.

The following measures and indicators are the most frequently used expressions of earning potential and profitability in the stock markets. They are also regarded most relevant and useful in other media, such as property, where a process of securitization is likely to reduce the difficulties of cross-media comparisons. To illustrate these measures and indicators an extract from the annual report of a hypothetical company True & Earnest plc is shown in Table 3.1.

Table 3.1 Extracts from the annual report of True & Earnest plc

Current prices: Ordinary share £ 0.75		
Loanstock (£ 100) £ 60		
Assets		£ 20 000 000
Equity capital		
£ 10 000 000 in ordinary (50p) shares		
£ 3 000 000 reserves		£ 13 000 000
Dept capital		
loanstock @ 8% to 1997		£ 6 000 000
Other liabilities		£ 1 000 000
Gross trading profit		£ 3 000 000
Interest on loan	£ 480 000	
Taxable profit		£ 2 520 000
Corporation tax	£ 1 260 000	
Equity earnings		£ 1 260 000
Dividend payout	£ 1 000 000	
Retained earnings		£ 260 000

3.3.1 Return on capital employed or return on investment (ROCE or ROI)

This is a fundamental measure of profitability. It is the ratio of the earnings and the capital committed to the enterprise. Although this measure is popular, nevertheless it is rather difficult to define and care must be taken when using it for comparisons:

$$\mathbf{ROCE}\ or\ \mathbf{ROI} = \frac{\text{Earnings}}{\text{Capital employed}}$$

The earnings or profits are usually pre-tax operating earnings or profits. In some cases, after-tax earnings may be more appropriate. The capital employed is the total of the long-term funds employed in the enterprise. This is usually the sum of the share capital and the debt capital. Further additions or deductions are sometimes appropriate and this aspect should be properly explained.

In the case of True & Earnest plc:

$$\mathbf{ROCE}\ or\ \mathbf{ROI} = \frac{£\ 3\,000\,000}{£\,20\,000\,000} \times 100\% = 15\%$$

3.3.2 Earnings per share (EPS)

This ratio is computed by dividing the number of ordinary shares into the earnings allocated to equity shareholders:

$$EPS = \frac{Equity\ earnings}{number\ of\ shares\ issued}$$

For True & Earnest plc:

$$EPS = \frac{£\ 1\ 260\ 000}{20\ 000\ 000} = 6.3p\ per\ share$$

3.3.3 Earnings yield

This is the ratio of the earnings per share and the current price of a share:

$$Earnings\ yield = \frac{EPS}{Current\ price\ of\ a\ share}$$

For True & Earnest plc:

$$Earnings\ yield = \frac{6.3}{75} \times 100\% = 8.4\%$$

3.3.4 Price/earnings (P/E) ratio

This ratio relates earning to the price of a share:

$$P/E\ ratio = \frac{Current\ price\ of\ a\ share}{Earnings\ per\ share}$$

For True & Earnest plc.

$$P/E\ ratio = \frac{75}{6.3} = 11.90$$

$$Dividend\ per\ share = \frac{Dividend\ payout}{Number\ of\ shares\ issued}$$

For True & Earnest plc:

$$= \frac{£\ 1\ 000\ 000}{20\ 000\ 000} = 5p\ per\ share$$

3.3.5 Dividend yield

This expresses the relationship of the current share price and the dividend earned by each share. It is usually computed on a 'grossed up' basis to facilitate direct yield comparisons with other investments.

$$Dividend\ yield = \frac{Dividend\ per\ share \times 100}{Current\ share\ price \times (100 - T)} \times 100\%$$

where T is the standard rate of income tax.

For True & Earnest plc:

$$\textbf{Dividend yield} = \frac{5 \times 100}{75 \times (100 - 27)} \times 100 = 9.13\%$$

3.3.6 Gearing ratios

Gearing is the ratio of fixed-interest capital (preference shares, debentures and other debts) to the total capital employed. It is usually expressed as the **ratio of debt to equity** and measured by long-term borrowing and the shareholder's funds in the balance sheets. Although gearing is usually considered as an indicator of risk, it does have a considerable impact on the returns to the investor.

Capital gearing is the ratio of debt capital to the total capital employed, expressed as a percentage. Higher capital gearing ratios usually indicate an increased sensitivity to asset value changes:

$$\textbf{Capital gearing} = \frac{\text{Debt capital}}{\text{Total capital}} \times 100\%$$

For True & Earnest plc:

$$\textbf{Capital gearing} = \frac{£ \ 6\,000\,000}{£\,19\,000\,000} \times 100 = 31.58\%$$

Income gearing is the ratio of the interest payments on the debt capital to the gross trading profit:

$$\textbf{Income gearing} = \frac{\text{Interest payments}}{\text{Gross trading profit}} \times 100\%$$

For True & Earnest plc:

$$\textbf{Income gearing} = \frac{£ \ 480\,000}{£\,3\,000\,000} \times 100 = 16\%$$

Warning: Most of the above ratios and indicators may be defined in a variety of different ways. Care must be taken with regards to the way the calculations were made in order to ensure like-with-like comparisons.

3.4 TOWARDS THE DEFINITION OF RISK

The prerequisite of the quantification, measurement and analysis of the effects of uncertainty is its definition in appropriate terms.

Investors inevitably encounter uncertainty which will surround the consequences of their actions. The main source of this uncertainty is time. The precise forecasting of future events is virtually impossible and the approximation of the future becomes more unreliable with the passage of time.

Uncertainty is synonymous with the lack of knowledge and information. On the basis of the amount of knowledge and information available, a **spectrum of uncertainty** can be identified. This spectrum ranges from Certainty into Total Uncertainty and may be subdivided into Risk and Partial Uncertainty. Figure 3.1 shows the spectrum of uncertainty, together with its subdivisions.

Certainty   Risk   Partial uncertainty   Total uncertainty

Fig. 3.1   The spectrum of uncertainty.

At one extreme of the spectrum lies Certainty, where there is a precise knowledge of the outcome.

Risk is a situation where alternative outcomes are identifiable, together with a definite statement of the probabilities of such outcomes.

In the case of Partial Uncertainty, alternative outcomes can be identified but without the knowledge of the probabilities of such outcomes.

Total Uncertainty means that neither alternative outcomes nor their probabilities can be identified.

In reality, the boundaries of these subdivisions are blurred. Nevertheless, the subdivisions are important as they enable the design of the most appropriate methodology to treat the various degrees of uncertainty affecting the investment decision-making process.

Most investment activity is carried out under the conditions of Partial Uncertainty. However, in most cases decision-making appears to be easier if conditions of risk are assumed.

Although the above discussion defines **Risk** as a distinct part of the spectrum of uncertainty, it is appropriate at this stage to consider other concepts of risk.

The term risk possesses a variety of meanings, in general it is used to denote the exposure to adversity or loss. In the context of investment, risk is usually used to describe the unpredictability of the financial consequences of actions and decisions. In business operations risk usually refers to a set of unwanted and uncertain events. Risk in the analytical sense will be regarded as the description of the extent to which the *actual outcome* of an action or decision may diverge from the *expected outcome*. An action or decision is described as risk-free when its consequences are known with certainty. Investors usually describe the outcomes in terms of **expected returns**. The actual returns can be greater or smaller than the expected returns. To a rational investor who is adverse to risk, the possible deviation from the expected outcome, resulting in a lower actual return, has a greater importance. The possibility of achieving a lower return than expected is usually termed as the **downside risk**.

The degree of variation of possible outcomes from a decision or action is usually associated with risk. If the range of possible outcomes is expected to be wide, then the action or decision is considered to be more risky than if the range of possible outcomes is expected to be narrower.

Variability of outcomes is used to describe the degree of risks associated with various investment activities. As it will be discussed further in another part of this chapter, variability can be measured in absolute or in relative terms.

The usual definitions of risk are either descriptive or analytical. The descriptive definitions of risk are related to the sources and elements of risk. These definitions are particularly useful for the classification of investments on the basis of the risks associated with them. They are also used in the determination of risk premia to compensate for risk taking.

The analytical definitions of investment risk are usually expressed in either probabilistic terms or in terms of variability. Some of these analytical definitions of risk are as follows:

- The probability of loss.
- The probability that the investor will not receive the expected or required rate of return.
- The deviation of realizations from expectations.
- The variance or volatility of returns.

3.5 ELEMENTS OF RISK

Investors usually perceive investment risk in terms of the following:

- Security of the capital invested.
- Security of the expected incomes.
- Liquidity.
- Problems of asset management.

The total risk associated with investments is made up of several individual components. These components are usually classified under the headings of **systematic** and **unsystematic or specific risks**:

$$\text{Total risk} = \text{Systematic risk} + \text{Unsystematic risk}$$

Systematic risk is caused by factors which affect all investments. Changes in the general economic and political environment are the causes of the systematic risks. Unsystematic or specific risks affect only a particular investment.

The investor has no control over systematic risks since he has no control over the economic or political situation. On the other hand, he has a limited control over his exposure to unsystematic or specific risks. He can investigate the companies and industries in which he is going to invest and he can make his decisions on the basis of the track record and prospects of these companies and their management.

Although the risk components can be individually identified, they are not independent of each other.

The components of the systematic risk are as follows:

(i) *Market risk* is related to the fluctuations in the market for a given investment.

(ii) *Cyclical risk* is related to the variations in the business cycle.

(iii) *Inflation or Purchasing Power risk* is related to the uncertainty of the future purchasing power of the returns produced by investments.

(iv) *Interest rate risk* is related to the variability of the interest rates. The rate of return on bonds and Treasury bills is uncertain because of the interest rate risk.

The components of the Unsystematic or Specific Risk are as follows:

(i) *Business risk* is related to the risk associated with a company's business operations. The factors that influence this risk are to do with the size of the company, product mix, competition and the general orientation of the management team in charge.

(ii) *Financial risk* is dependent on the way the company's operations are financed. The larger the debt finance used by the company, the larger is the associated financial risk.

(iii) *Liquidity risk* encompasses the following aspects: the degree of difficulty associated with the realization of the capital invested; the divisibility and marketability of the asset; and the costs involved with the realization of the capital.

(iv) *Other specific risks* which usually affect individual investment are industry risk, location, etc.

3.6 THE QUANTIFICATION AND MEASUREMENT OF RISK

Any meaningful analysis of risk must be preceeded with its quantification and measurement. Quantifying risk is not an easy task as there are still serious arguments about the proper definition of risk. Most economists have identified investment risk with the dispersion of returns and the consensus is to relate risk to the variability of returns.

The concept of risk in the context of investment in a financial asset or security implies that the **actual rate of return** may well diverge from the **expected rate of return**. The investor estimates the probability distribution of the **future rate of return** about its *expected value*. Risk can therefore be defined and quantified as the amount of potential variability of the future rate of return about its mean or expected value.

The dispersion or variability of the rates of return can be measured in *absolute terms* in a number of different ways. The range of possible rates of returns can be computed by subtracting the smallest possible rate of return from the largest possible rate of return. Knowing the range of possible returns is better than having no estimates of dispersion, but it is not a satisfactory measure of risk. It says nothing about the probabilities of the various possible intermediate rates of return. Two investment propositions with the same range of possible rates of return could have widely differing variations around the mean of their respective distributions. If the distribution is subdivided into a number of **fractiles**, and the **inter-fractile ranges** are calculated, then the precision of the information provided by the range could be improved. If the distribution of the rates of return does not have a finite variance, then the only (albeit crude) measure of dispersion is the range and its refinements.

The most often used measures of variability are the **variance** and the **standard deviation**. The variance is the average of the squared deviations from the *mean* of a distribution, whilst the standard deviation is the square-root of the variance. When the distribution of the rates of return is *normal* (i.e. symmetrical and bell-shaped), then these two measures will describe the dispersion precisely. The justification to use the variance and standard deviation as an appropriate measure of risk is not restricted to normal distributions. As long as the shape of the distribution of the rates of return is symmetrical, with a finite variance, and the investor's risk preferences can be described with a quadratic utility function, the variance and the standard deviation can be regarded as the appropriate measures of risk.

The **mean absolute deviation** of the rates of return is also used as a measure of variability and regarded as an appropriate risk measure. It differs from the standard deviation, in that it gives less weight to large fluctuations since the deviations are not squared. Some analysts prefer this measure as it remains more stable through time.

The variance and the standard deviation cover both the desirable (above the mean) and the undesirable (below the mean) parts of the distribution of the rates of return. This fact is regarded by some people as a source of confusion as the investor would mainly worry about the likelihood and size of the undesirable elements in the distribution. To overcome this difficulty they recommend the use of the **semi-variance** and the **semi-standard deviation**. The semi-standard deviation is defined as the square-root of the expected value of the squared *adverse returns*, the difference between a minimum desired return and the actual return. The minimum desired return is either a computed target return or the return produced by an alternative investment, or some other chosen standard.

All of the above measures of risk rest on the assumption that the investor could somehow perceive the distribution of the future rates of return. This is the point where most of the difficulties of the quantification and measurement of investment risk emerge. Risk and uncertainty are associated with events in the future and, therefore, to attempt to measure risk *objectively* is a contradiction in terms. On the other hand, the assumption that the past is a useful and reliable base on which a reasonably accurate view of the future could be formulated seems to be valid albeit in the short run. So, by using the evidence from the past and the present prevailing trends, the investor should be able to perceive the shape of the distribution of future rates of return.

By using the rates of return figures achieved in the past, a **frequency distribution** can be drawn up and the appropriate statistics, the mean, variance and standard deviation, could be computed. On the very short run, it could be assumed that the future will not be significantly different from the past and the shape of the frequency distribution will be very similar to the probability distribution of the future rates of return.

However, the precision of the above extrapolation will deteriorate very rapidly and the drawing up of probability distributions will rely on reasoned, *subjective* inputs. Although there are many problems associated with the use of subjective judgements and individual perceptions in the decision-making process, there are a number of reliable methods available to draw up probability distributions on a subjective basis (the discussion of these methods are outside the scope of this chapter). Here it is sufficient to say that with a little care and thought the beliefs of individuals or groups concerning future outcomes can be converted into very useful probability distributions.

Having seen some of the most important measures of risk, based on the variability of the future rates of return, let us examine these concepts in more detail.

Expected value E(r): is the mean of the probability distribution of the potential future rate of return. It is calculated by summing each of the possible rates of return multiplied by its respective probability; thus:

$$E(r) = \sum_{j=1}^{n} p(r_j) r_j$$

Example:

Probability $p(r_j)$	Future rate of return r_j	$p(r_j) r_j$
0.1	10%	1.0
0.2	11%	2.2
0.25	12%	3.0
0.2	13%	2.6
0.15	14%	2.1
0.1	15%	1.5
1.0		$E(r) =$ 12.4%

Risk: is the dispersion or variability of the future rates of return about its mean – i.e. the expected value. It is calculated as the weighted average of the squared deviations of the possible rates of return and the expected value. The weights used are the probabilities of the possible rates of return. Thus:

$$\text{var}(r) = \delta^2 = \sum_{j=1}^{n} p(r_j)\,(r_j - E(r))^2$$

Example:

Probability $p(r_j)$	Rate of return r_j	Deviation $(r_j - E(r))$	$(r_j - E(r))^2$	$p(r_j)\,(r_j - E(x))^2$
0.1	10%	– 2.4	5.76	0.576
0.2	11%	– 1.4	1.96	0.392
0.25	12%	– 0.4	0.16	0.04
0.2	13%	0.6	0.36	0.072
0.15	14%	1.6	2.56	0.384
0.1	15%	2.6	6.76	0.676
1.0				

Variance var(r) = 2.14

Standard deviation = 1.46

The above measures of risk treat the high values of the possible future rates of return in the same way as the possible low values of return. This symmetrical treatment of high and low values can lead to some confusion when investment propositions are ranked on the basis of their riskiness. These problems usually surface if the distributions are not symmetrical but skewed and the investor is particularly interested in the probability of not achieving a certain target rate of return. In these situations, the concept of the downside risk may be more appropriate.

Let us examine two investment propositions, Project A and Project B. The probability distributions of possible future rates of return are outlined in Fig. 3.2. Assuming that the investor is particularly concerned about the probability of not achieving a specified target rate of return R_T. The question is: which proposition is the more risky with regard to the investor's objective?

At first glance, the distribution of Project A appears to suggest less risk as it is much narrower and well peaked. The distribution of Project B is a much flatter and wider shaped, indicating a greater degree of variability about the mean of that distribution. However, the investor is mainly worried about the probability of the project not achieving the minimum acceptable target rate of return. So which one of the two projects has the greater probability, or risk, of not reaching the target set?

The probability of not achieving the target is directly proportional to the area under the respective distribution curve up to the vertical line erected at the point representing the target rate R_T.

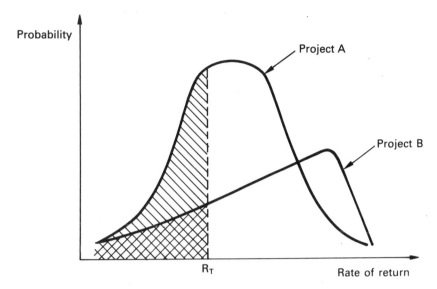

Fig. 3.2   Downside risk.

Referring to Fig. 3.2, Project A is more risky as the area under the distribution curve and the line representing the target rate is greater than the area under the distribution curve of Project B.

Clearly, this measure of downside risk has a great intuitive appeal, since it measures risk in terms of probabilities, which corresponds more closely to the general perception of risk. The main drawback of using downside risk as a risk measure in analytical work is that it is very cumbersome in computational procedures.

The absolute measures of variability, the variance and standard deviation of the distribution of future rates of return do not give the complete picture of the risk characteristics of investment assets. These measures of risk do not reflect the likely behaviour of assets *relative* to the market in which they are traded. An appropriate measure is required to reflect the variability of the rates of asset returns in the light of the changes occurring in the market. This new concept, usually referred to as **volatility**, is to provide the basis for the prediction of the future rates of asset returns under different market conditions.

The relationship between risk and return in the context of a particular market can be established by using a simple **regression technique**. By plotting the expected return on a particular asset against some suitable indicator of the state of the market, a scatter diagram can be produced which would reflect the relationship between asset returns and changes in the market. This relationship can then be formalized by establishing the line of best fit; Fig. 3.3 illustrates this process.

The equation of the regression line is as follows:

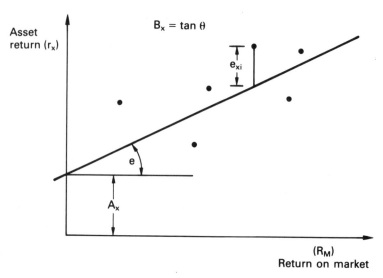

Fig. 3.3   Scatter diagram – line of best fit.

$$r_X = A_X + B_X \cdot R_M + e$$

where r_X is the return on a particular asset or security (x);

R_M is the return on the market expressed in terms of some suitable market index;

A_X is the intercept of the regression line on the vertical (return) axis;

B_X is the slope of the regression line and also the measure of the rate of volatility of the return on the asset relative to market changes;

e is an error term, showing the difference between an r and its value predicted by the regression line.

The return on the market is usually computed as the rate of change of a market index such as the FT All Shares Index, Dow-Jones Index or, in the case of the property market, an index like the JLW Property Index.

The coefficient B_X expresses the responsiveness or *volatility* of the expected asset returns to the changes in the market. B_X, referred to as the **beta coefficient**, is the measure of the systematic risk of an asset or security. If the numerical value of B_X is equal to 1.0, then the asset returns will move in unison with the changes in the market. If B is less than 1.0, then the asset returns will exhibit some sluggishness in following the changes in the market, whilst B_X values greater than 1.0 will mean an increase in volatility.

As seen above, there are a number of different measures of risk available. However there is no standard measure, although variance and standard deviation seem to be favoured by analysts because of their relative computational simplicity.

As a summary to this section, the list of the most frequently used quantitative risk measures are given below:

- Variance
- Standard deviation
- Mean absolute deviation
- Downside risk
- Beta coefficient
- Semi-variance
- Semi-standard deviation
- Range
- Inter-fractile range

3.7 SUMMARY

Since the prime objective of all rational investors is the achievement of adequate and satisfactory returns, the measurement of returns in the past and the assessment of possible returns in the future will always be an important topic. Earnings or returns are the sum of the returns from incomes, such as dividends, interest, rents, etc., and of the returns from the changes in the value of investment assets.

There is a variety of measures and indicators of returns and an alarming variety of definitions and computational procedures. It is very easy to get confused and great care must be taken when interpreting the computed results.

Investors need to look further than the plain figures of the rates of return or other indicators of earnings and profitability and consider cash flow profiles, timings of large capital injections, re-investments, gearing and many other aspects. It is virtually impossible to compress the effects of such a multitude of factors into a single figure.

Over short periods rates of return and other indicators tend to fluctuate widely, whilst over longer terms these measures tend to converge.

The definition, quantification, measurement and effective treatment of risk represent the most challenging aspect of investment decision-making. The effects of risk and uncertainty cannot be avoided in any investment activity. Investors are well aware of the dangers in trying to ignore these possible perils, although they do tend to get frustrated with the slow progress in the development of a suitable and effective methodology.

The definition of investment risk and its components is still subject to debate. There are a number of verbal definitions, quite a number of which are unsuitable for serious analytical work and defy all attempts of quantification and measurement.

It is now generally accepted that uncertainty and unpredictability are one and the same thing; and as unpredictability is usually associated with variability, the consensus is to define and measure risk and uncertainty with some measure of variability. There are a variety of measures of variability available for this purpose, each with some attractive and some awkward features. However, the variance and standard deviation seem to emerge as the most popular, absolute measures of variability, most suitable to measure risk.

A highly important development was the partitioning of the total investment risk into the systematic or market-related risk, and into the unsystematic or specific risk. The systematic risk required an appropriate relative measure of variability: volatility. This development has also had a profound effect on the treatment and management of risk.

A number of different approaches have also emerged for taking risk into account when investment decisions are to be made or when the monitoring of investment performance and efficiency are the main objectives. There is a promising progression in the development of such approaches. Although the early, traditional approaches are still favoured by many, the more sophisticated probabilistic approaches are now gaining ground. With the advent of abundant and relatively cheap computer power, the simulation of risky situations will give the much desired means to assess risk.

FURTHER READING

The following is a small fraction of the published work on this topic. The references are given here in order of ease of reading.

Lumby, S. (1981) *Investment Appraisal and Related Decisions*, Nelson, London.

Byrne, P. and Cadman, D. (1984) *Risk, Uncertainty and Decision-making in Property Development*, E. and F.N. Spon, London.

Hull, J.C. (1980) *The Evaluation of Risk in Business Investment*, Pergamon, Oxford.

Coyle, R.R. (1972) *Decision Analysis*, Nelson, London.

Sharpe, W.F. (1977) *Investments*, Prentice-Hall, Englewood Cliffs, NJ.

Bank Administration Institute (1968) *Measuring the Investment Performance of Pension Funds for the Purpose of Inter-fund Comparisons*, BAI, Park Ridge, Illinois, USA.

Hertz, D.B. (1964) Risk Analysis in Capital Investment, *Harvard Business Review*, January–February, 95–106.

Robicheck, A. (1975) Interpreting the results of risk analysis, *Journal of Finance*, **XXX** (5), 1384–6.

Markowitz, H. (1959) *Portfolio Selection: Efficient Diversification of Investments*, Wiley, New York and London.

Sykes, S.G. (1983) The assessment of property investment risk, *Journal of Valuation*, **1**, (3), 253–67.

Firth, M. (1975) *Investment Analysis*, Harper and Row

Levy, H. and Sarnat, M. (1982) *Capital Investment and Financial Decisions*, Prentice-Hall, Englewood Cliffs, NJ.

Society of Investment Analysts (1974) *The Measurement of Portfolio Performance for Pension Funds*, SIA, London.

Warren, R. (1983) *How to Understand and Use Company Accounts*, Business Books, London.

Articles

Hall, P. and Hargitay, S. (1984) Property portfolio performance – a selected approach, *Property Management*, 2, 218–29.

Hetherington, J. (1980) Money and time weighted rates of return, *Estates Gazette*, 256, 1164–65.

Messner, D. and Chapman Findlay, M. (1975) Real estate investment analysis: IRR versus FMRR, *Real Estate Appraiser*, July–August, **40**(4).

Newell, M. (1985) The rate of return as a measure of performance, *Journal of Valuation*, 4, Autumn, 130–42.

4	# Investment performance

4.1 INTRODUCTION

Investment decisions are usually made on the basis of beliefs about the future. Such beliefs are based on past experience. Most investors accept the fact that their beliefs about the future are imperfect and all their investment activities will, inevitably, involve some risks. All rational investors aim to achieve maximum returns whilst trying to minimize risks. Investment performance and efficiency are the expressions of the degree of achievement of that aim.

The principal objective of an investment manager's work is the creation and maintenance of successful and efficient portfolios of investment assets. The *raison d'être* of any investment portfolio is the promise of greater investment efficiency in terms of the continuing dynamic interplay between return and risk. It is essential therefore that methods through which the success and efficiency of investment portfolios can be assessed, are devised and put into practice. Only through the continuous monitoring of the achieved results can investment strategies succeed.

Investment performance, its measurement and analysis, have attracted considerable attention over the past thirty years. Attention has only recently turned towards the need for a reliable measurement of the investment performance of property assets.

The large investment portfolios of various mutual funds, pension funds, insurance companies and other financial institutions contain substantial holdings of assets other than stock market securities such as property, works of art, etc. The trustees of such large funds realize the need for the measurement of historic performance and the assessment of the likely future performance of their investment portfolios. Through the monitoring and analysis of portfolio performance the investor can gain valuable insight into the investment characteristics and behaviour of the various assets included in their portfolios. Such characteristics and behaviour are viewed in the context of the movements of the various sectors of the investment market. Only through a standard method of performance measurement can the true contribution, good or bad, of the different assets and sub-portfolios to the general investment portfolio be evaluated.

Until quite recently, property investment performance has been considered too complex and too specialized to be analysed side by side with other investment assets. The main reason for this attitude is to be found in the special investment characteristics of property and of the property investment market. Risk measurement in the property medium is even more complex than in other investment media. These are the reasons why the appropriate measurement methodology for property portfolio performance is still in its early stages of development.

Performance may be defined as achievement, relative to targets and objectives. By measuring performance, the degree of achievement against a set of objectives and targets can be expressed in quantitative terms. The shortfall, or excess, relative to the targets can then be analysed and useful conclusions and explanations be drawn for decision-making.

Performance analysis is a vital component of the decision-making process. Rational decision-making at all levels would be virtually impossible without the quantified evidence of past performance and a reasoned assessment of likely future performance.

The investor or the trustees need to satisfy themselves that the invested funds produced returns which satisfy the targets and objectives of their investment strategy and that future prospects justify lowering, maintaining or increasing the level of investment in certain assets or portfolios. They are also interested in the appraisal of the success and effectiveness of the management of assets and portfolios.

The investor seeks answers to a variety of questions, some of which relate to their general investment portfolio, whilst others seek answers and explanations regarding the contribution of the sub-sets, such as property portfolios, to the performance of the general portfolio.

The most frequently posed questions are:

- How have the individual investment assets performed in absolute terms and relative to some chosen yardstick?
- How do the various sub-sets of the general portfolio, such as the Equity portfolio, Gilt portfolio and, say, the Property portfolio, compare?

- What are the principal factors which affect the performance of the assets in the various media?
- What are the risks associated with the various media and how can these risks be managed?
- Can further investment in a particular medium be justified?
- When, how much and in what type of asset should further capital investment be made?
- Should the existing general portfolio or the sub-portfolios be restructured?

Until recently, the measurement and assessment of investment performance was regarded only from the point of view of the investor or the trustees. It has been recognized, however, that fund and portfolio managers also require the benefits of performance measurement and assessment for internal use. On the operational level, the benefits of performance analysis are considerable. In the monitoring of the achievement of the targets set by the investor for the portfolio manager, performance analysis is more useful than any other analytical decision-making tool.

It should be stressed, however, that the measurement and analysis of investment performance will never provide totally reliable and precise answers. The fundamental problems of risk and the subjectivity of asset valuation procedures are the reasons for this.

4.2 THE PURPOSE AND OBJECTIVES OF PERFORMANCE APPRAISAL

Those interested in the appraisal of investment performance are trustees and managers of pension funds, managers of unit trusts and insurance companies, trustees of charities and friendly societies and the private investor.

Whilst institutional investors and trustees are responsible and accountable for the stewardship of other people's money, they need to know what is happening to the invested funds. Private investors, on the other hand, are interested in the income obtained (and obtainable) from the investment or the amount of capital growth, or a combination of these. The private investor is usually the owner of unit trusts units or investment trusts shares and he is interested in the comparison between the performance of his units and shares, with that of other unit trusts and investment funds.

Institutional investors and trustees require performance appraisal for the following four reasons:

1. *To monitor the progress of invested capital.* This monitoring is absolutely essential as the investments are usually held to meet certain liabilities in the future. They must make sure that the progress of the investment assets will ensure that these liabilities can be met when they fall due.
2. *To monitor relative performance.* Institutional investors do like to pitch their expertise against the expertise of the managers of other funds, they

are particularly interested in how well, or how badly, their investment activities fared in this respect. The fund's performance is compared with the performance of the investment markets and also against the ups and downs of the economy, as reflected in the general economic indicators.

3. *To analyse past performance for decision-making.* The allocation of funds to new investment activity is usually done on the basis of the analysis of past performance. Institutional investors disaggregate the historic overall return into two components: a component due to investment strategy and a component due to the returns of the individual assets and securities. From the analysis the relative weight of strategic elements (i.e. the proportions held in various asset groups) and the relative weight of security and asset returns will emerge. Decisions regarding the correction of imbalances can then be made.

4. *To evaluate the performance of management.* Performance measurement will provide an insight into the effectiveness and efficiency of the investment managers. This is particularly important if the responsibility of fund management is farmed out to external investment managers and analysts.

The results and conclusions of performance appraisal are summarized in a performance report. Such a performance report is expected:

(i) to quantify historic performance and measure it against some chosen standard;
(ii) to provide explanations for good or bad performance;
(iii) to assess in quantitative terms the expected future performance to see if the prospective performance is likely to meet the targets set;
(iv) to assist in the re-assessment of investment strategies and to point to possible adjustments.

On the other hand, on the *operational* level the portfolio manager needs a tool with which he can evaluate the performance of the investment asset in his care, and also to analyse the effectiveness of his own decision-making in retrospection. The portfolio manager's objectives in analysing investment performance are as follows:

1. To monitor the degree of achievement of the targets set in terms of returns and risks.
2. To draw conclusions from the analysis of historic performance figures.
3. To plan future action in order to achieve maximum returns at an acceptable level of risk by making the appropriate adjustments to his portfolio through acquisitions and disposals.

4.3 MODES OF PERFORMANCE APPRAISAL

There are two principal modes of performance appraisal: the first is the historic or **retrospective mode**, and the second is the **prospective mode**.

The measures and indicators used in both modes are the same, but whilst the historic measures can be objectively determined, the indicators of prospective performance can only be assessed on a subjective basis.

The appraisal of investment performance may be carried out in absolute terms and/or relative to the performance of other portfolios or investment opportunities.

It must be stressed, that the measurement of investment performance without subsequent analysis does not have any virtue as far as decision-making is concerned.

4.4 THE MEASURES AND INDICATORS OF PERFORMANCE

The measures and indicators of performance are expected to express, in quantitative terms, the achievement of targets and objectives in the context of the market which generates the risk–return relationships, or trade-offs.

The rate of return is regarded as the most important measure of performance.

The performance of a portfolio or fund is usually interpreted in a broader sense than the rate of return. The other important element of performance is the risk associated with the portfolio or the fund. The risk element can be defined and measured in a number of different ways. The different aspects of risk can be reflected and assessed in terms of the following:

1. The *variability* of the rate of return on the portfolio.
2. The *volatility* of the rate of return on the portfolio which indicates its sensitivity to the rate of return on the market.
3. The *diversification* of the portfolio expressed by the portfolio balance.
4. The *downside risk* of the portfolio, which is the probability that a specified target rate of return will not be achieved.

All these elements of portfolio performance need to be appraised and analysed if a clear picture of the performance of the portfolio is to be presented and when a strong base needs to be established for effective decision-making.

4.4.1 The rate of return

The rate of return can be calculated in a number of different ways. The appropriate method of calculation depends on whether the rate of return is required to measure the performance of the portfolio as an aggregate of the returns on individual assets, or it may be required to measure the portfolio manager's performance.

The definitions and computations of the various rate of returns are fully discussed in Chapter 12.

4.4.2 Risk measures and indicators

The appropriate risk indicators have been discussed in Chapter 3, and therefore the risk indicators used in performance appraisal will be only briefly discussed here. The risk dimension of investment performance is treated on two different levels in portfolio performance appraisal.

The efficiency of the management of the risk dimension is evaluated by examining the diversification of the portfolio. To reflect the level of diversification, the **portfolio balance** is used. The portfolio balance simply expresses the proportions of the components of the portfolio according to a particular definition. Portfolios can be diversified in a number of different ways. For example, a property portfolio may be diversified on the basis of asset classes (offices, shops, industrials, etc.) and on a geographical basis (London, Southeast, South, Midlands, etc.). The portfolio balance reflects diversification on these different levels, indicating the degree of elimination of the **unsystematic component** of the total risk of the portfolio.

Other risk indicators used in performance measurement are the *variance* of the rate of return on the portfolio, whilst the volatility of the portfolio return is measured by the portfolio beta.

There are other risk indicators in use, reflecting the special risk characteristics of specialized portfolios. Such special risk indicators will be further discussed in the context of the performance measurement of property portfolios.

The **downside risk** is often used in the prospective mode of performance appraisal. This particular risk indicator is most appropriate when a simulation approach is used to assess the future performance of the portfolio.

4.4.3 Other indicators

Investors may be interested in aspects of investment performance other than the rate of return and the expression of risk. They are often concerned about **rate of growth in capital values** and in **rate of growth of incomes**.

There is also a need for appropriate bench marks for comparisons. Various market and economic indices are used for the purpose.

4.5 THE STAGES OF THE APPRAISAL OF PORTFOLIO PERFORMANCE

The portfolio is regarded as an integrated system of assets whose *raison d'être* is to enable the investor to achieve his investment objectives with maximum efficiency. An efficient portfolio will produce maximum returns at an appropriate minimum level of risk. The performance of the portfolio is therefore entirely dependent on the efficiency of the system as a whole. Portfolio performance is not a simple sum of the performances of its parts, as the way in which the individual components fit into the system is also of great importance. Thus the ideal appraisal of investment performance proceeds from the performance appraisal of the components towards the examination of the performance of the portfolio, treated as a single entity, as an intergrated system. This is usually a three-stage process:

1. The appraisal of the performance of individual assets.
2. The appraisal of the performance of the various asset sectors.
3. The appraisal of the performance of the portfolio.

4.6 APPROACHES TO PERFORMANCE APPRAISAL

During the past four decades a number of different approaches to the appraisal of investment performance have evolved. The evolution of these approaches is parallel with the development of the theoretical framework explaining the dynamic interrelationships in the world of finance and investment. In order to understand the rationale of the various approaches the basic requirements need to be established.

It should be noted that the basic investment vehicle is the individual asset and the more sophisticated investment vehicle, the investment portfolio, is an attempt to raise the efficiency of the overall investment activity. At the base level, the performance of the individual asset is seen in terms of the rate of return produced, or expected, and the risk associated with that asset. The risk is measured in terms of the variability of historic returns and the probability distributions of prospective returns. Usually performance appraisal is carried out on portfolios and investment funds by accepting that the performance of these investment vehicles will, ultimately, depend on the performance of the individual assets contained therein.

The various approaches to performance measurement will now be described in the sections below. Most of them are **comparative approaches**, only a few approaches express investment performance in absolute terms.

4.6.1 The early approach to performance appraisal

The original test of performance was the calculation of the yield of a portfolio or fund. The calculation was based on the following formula:

$$\textbf{Yield} = \frac{2 \times (\text{Income received during the year})}{V_B + V_E + (\text{Income received})} \times 100\%$$

where V_B is the book value of the portfolio at the beginning of the year; and V_E is the book value of the portfolio at the end of the year.

In those early days, the portfolios consisted mainly of fixed interest stocks and mortgages, and only a very small proportion of ordinary shares, and therefore the formula gave a reasonable account of the return on the invested funds. The main attraction of this simple approach was that it required no more information than that found in published accounts.

This approach is no longer adequate as capital appreciation is a major component of the total return and the proportion of ordinary shares has increased manifold. The approaches to performance appraisal fall into two categories:

1. Approaches derived directly from Portfolio Theory.
2. Pragmatic approaches.

4.6.2 Approaches derived from portfolio theory

These approaches are based on the various portfolio selection models developed by Harry Markowitz, William Sharpe and others in the USA. The appeal of these approaches is that they incorporate both the rate of return and the risk of the portfolio. The principal criticism of these methods is their reliance on the beta coefficient as a representation of investment risk, as the beta coefficient is not constant over time.

Performance appraisal using the Markowitz model

Using the Markowitz model, the frontier of efficient portfolios can be determined for a particular period. The rate of return on the portfolio or fund can then be compared with that of the optimum portfolio which would have the same level of risk as the portfolio or fund. Theoretically this is the best approach to performance appraisal. This method suffers from the major disadvantage of extreme computational complexity and is rejected by most practitioners who cannot accept the variance and standard deviation as the only measure of risk.

Performance appraisal using Sharpe's 'reward to variability ratio'

This approach is based on the 'characteristic line' which can be established by the regression between the rate of return on the portfolio and the rate of return on the market.

The procedure is rank portfolio performance on the basis of the standard deviation of historical returns. The reward to variability ratio is computed as follows:

$$\textbf{Reward to variability ratio} = \frac{(R_a - R_f)}{\delta}$$

where R_a is the average rate of return;
R_f is the risk free rate of return;
δ is the standard deviation of the portfolio's return.

The higher the value of the reward to variability ratio, the better is the portfolio's performance for a given level of variability (risk).

Performance appraisal using Treynor's method

The rate of return on the portfolio or fund is plotted against the rate of return on the market over the same period. From a series of such points the 'characteristic line' can be drawn for a portfolio or fund. The return on the market is represented by the change in a market index such as the FT All Shares Index.

Figure 4.1 illustrates the comparison of the performance of two funds, Fund A and Fund B. Assuming a risk-free rate of return, say, index-linked gilts, of 8%, Fund A exhibits superiority over Fund B as the line representing the risk-free investment cuts the line of Fund A before it cuts the line of Fund B.

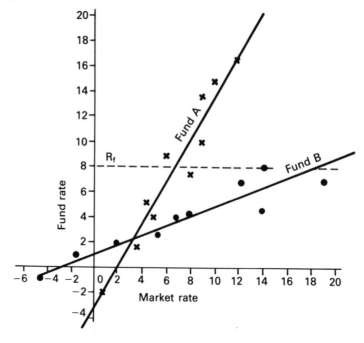

Fig. 4.1 Comparison of investment performance.

4.6.3 Pragmatic approaches to performance appraisal

These approaches attempt to bypass the theoretical and computational diffi-culties of the methods mentioned above and simplify procedures by using suitable benchmarks against which the performance of portfolios could be measured and assessed.

Performance appraisal using 'random portfolios'

This method has gained popularity in the USA for the performance appraisal of investment funds which are almost entirely invested in stock market securities. The basis of the method is to use the performance of randomly selected portfolios as the standard of comparison for measuring portfolio performance. For this purpose, a number of random portfolios are selected, all about the same size as the portfolio whose performance is to be measured. The supporters of this method argue that a **managed portfolio** should perform better than the average random portfolio, and from the comparison of performances, a measure of the effectiveness of the management should emerge.

The main disadvantages of this approach are the dependence of the result on chance and that it ignores the constraints and restrictions imposed on the managers.

Performance appraisal using 'control funds'

The rationale of this approach is to set up a hypothetical fund, which is invested in a fixed composition of securities, to represent the investment market. Such a fund is usually referred to as **control** or **notional fund**. The procedure is to match every transaction in the real fund with an identical transaction in the notional fund. The transactions in the notional fund are the hypothetical purchase or sale of units in an appropriate index.

This approach does incorporate into the appraisal the timing and size of cash flows. Through the comparison of the performances of the real fund and of the notional fund, the effectiveness of the portfolio manager's decision-making can be assessed.

Performance appraisal by direct comparison with 'market performance' or with the 'economy'

Probably the simplest and most straightforward approach to the monitoring of portfolio performance is the direct comparison of the portfolio returns with those of the investment market, or the economy. The method assumes that the investor's portfolio was assembled with a definite strategy in mind. The strategy would reflect the investor's risk posture, objectives and targets, both in the short and in the long run. Depending on the required precision of the comparisons, the investor would choose a suitable time interval over which the rate of return comparisons were to be made. The successive rate of return comparisons show not only the superiority or inferiority of portfolio returns, but also give an insight into the risk characteristics (variability and volatility) of the portfolio. Decisions can be made about adjustments on the basis of the evidence emerging. Figure 4.2 illustrates the rationale of this approach.

This approach is very attractive because of its simplicity. Care should be taken when choosing the index for the comparisons. There are a number of

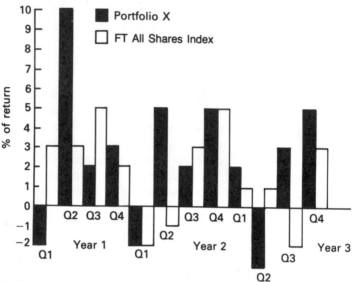

Fig. 4.2 Comparison of quarterly returns on portfolio X with quarterly returns on the FT All Shares Index.

market indices available, each intended to reflect a particular aspect of the investment market. Chapter 3 contains the description of these indices; a shortlist of these indices will suffice here:

- FT Industrial Ordinary Index
- FT 500 Share Index
- FT All Shares Index
- S&P 500 Index
- FT Actuaries 15 Year Gilt Index
- JLW Property Index
- IPD Monthly Index

The Retail Price Index is the most frequently used indicator of the economy in the appraisal of portfolio performance.

The direct comparison approach is particularly useful for those funds and portfolios which are invested in liquid assets as adjustments can be made to the portfolio without delay.

League tables

Several leading firms of consulting actuaries and investment advisers offer performance comparison services in the form of **league tables**. Investment funds, particularly unit trust funds, compete for the flow of investments into their funds. The league tables provide the ranking of investment funds in the order of their achievements, which are attributed to the expert management of the fund.

Potential investors can draw conclusions from the published league tables and decide where to channel their investment monies.

League tables have a great potential for misuse and misinterpretation and they have little value in serious decision-making. The main drawback of the league table approach is that the rankings are made without taking into account the objectives, risk characteristics and sizes of the funds participating.

The 'rational' approach to portfolio performance appraisal

Most of the large investors now have the capability to maintain databases containing the historic record of their investment activities and also have access to market and economic information on a large scale. They also have access to relatively cheap computer power. With all these facilities available, the management of their investment portfolios can now be organized on a rational and systematic basis. Performance appraisal is an integral part of the systematic management approach and there is no reason not to carry out performance appraisal and monitoring in-house.

In-house performance appraisals should employ all the pertinent measures and indicators, discussed in Section 4.5. In order to guard against the deluge of information, the information flow resulting from the performance appraisal should be carefully rationed to those who can utilize the information without delay. The periodic **performance report** should be concise, highlighting both successes and shortcomings. Comparisons with external benchmarks and internally set targets should also be included.

The simulation approach to the appraisal of portfolio performance

This approach is particularly useful for the assessment of future performance. Portfolio returns are computed using discounted cash flow models. The determinants of future cash flows are imputed in terms of their probability distributions and the simulations are carried out using the Monte Carlo process. The output of the simulations is the probability distribution of the **expected rate of return** on the portfolio.

The risk characteristics can also be assessed from the probability distribution of the return. Either the usual measures of **variability** or the **downside risk** can be used to express the portfolio risk in quantitative terms.

This approach depends on the availability of a suitable computer and appropriate software.

4.7 THE PROBLEMS IN PERFORMANCE APPRAISAL

There are a number of serious problems in the appraisal of investment performance. The first and most difficult problem is associated with the subjectivity in the valuation of assets. The capital values assessed from various valuation procedures are a vital part of the computation of the rate of return.

The presence of **investment risk** means that none of the evaluation methods are precise, as most measures of investment risk are still awaiting universal acceptance.

Most of the other problems affect the comparative appraisal of investment performance.

There is a particular problem of accounting for the inflow of new monies. There is a serious loss of precision if the returns are not measured each time new monies flow into the fund or portfolio.

When comparing different funds, the dividend incomes and other internally generated money flows may occur at significantly different times, and similarly, monies are withdrawn from portfolios at different dates. These money movements can cause significant distortions, rendering comparisons invalid.

Various funds and portfolios often have different constraints and limitations imposed on them. Such limitations can be statutory or can be operating constraints. For example, unit trusts usually prefer to invest in liquid assets, whilst charitable trusts have often had restrictions on the way the income is to be spent for charitable purposes. The relative evaluation of constrained funds can only be attempted if the comparison is done with similarly constrained portfolios or funds. The comparison of the performance of constrained funds with market indices is inappropriate.

There are also serious problems associated with the performance appraisal of mixed-asset portfolios containing a number of different assets from a number of investment media. Such assets exhibit significantly different investment characteristics in terms of the rate of return and risk. The presence of property assets in investment portfolios create extreme difficulties as the performance appraisal of property assets require special treatment. The best way to resolve the problem of mixed-asset portfolios is to split them into sub-portfolios, representing the different investment media. Then evaluate the performance of each of the sub-portfolios separately and attempt to aggregate their performance contributions into the performance appraisal of the general portfolio.

4.8 SUMMARY

Investors attempt to measure the performance of their investment assets and portfolios to make certain that their investment objectives are being met. Performance appraisal is one of the most important tools of the effective management of investment assets. Performance appraisal will help in the identification of the weaknesses in the investment portfolios and can also be used to reveal the strength and weaknesses of the professional management of investment assets.

As far as investment decision-making is concerned, performance measurement helps to identify the areas where appropriate adjustments need to be made – i.e. acquisitions and disposals – in order to redress any imbalance in the portfolio. The timing of investment activity can also be improved in the light of the evidence provided by performance appraisal.

If performance appraisal is to be effective and reliable, then it must be approached systematically. The measures and indicators of performance must be clearly defined and understood. The investor must decide which of the approaches would be most appropriate for his purposes. Although performance appraisal involves a lot of computations and requires the availability of a considerable amount of data, computer technology is now more than adequate to cope with the computational loads.

The computerized analysis of the performance of investment assets and portfolios can produce an alarming array of measures and indicators. The analysis and explanations of such indicators require commonsense and, quite often, special expertise. Because of the presence and involvement of risk and uncertainty and a considerable degree of subjectivity, the whole process can never be regarded as absolutely 'correct' and totally reliable.

FURTHER READING

The most recent and comprehensive work published on this topic is:

Hymans, C. and Mulligan, J. (1980) *The Measurement of Portfolio Performance*, Kluwer, London.

The following is a small selection from the published books and articles on this topic:

Bank Administration Institute (1968) *Measuring the Investment Performance of Pension Funds*, BAI, Park Ridge, Illinois, London.
Society of Investment Analysts (1972) *Portfolio Performance Measurement for Pension Funds*, SIA, London.
Cocks, G. (1972) An objective approach to the analysis of portfolio performance, *Investment Analyst*, **34**, December, 3–7.
Dietz, P.O. (1966) *Measuring Investment Performance*, Columbia University Press, New York.
Eadie, D. (1973) A practical approach to the measurement and analysis of investment performance, *Investment Analyst*, **37**, December, 12–18.
Gilland, A.B. (1962) Measuring ordinary share portfolio performance, *Investment Analyst*, **3**, August, 30–5.
Treynor, J.L. (1965) How to rate management of Investment Funds, *Harvard Business Review*, **43**, (1), January–February, 63–76.

Quantitative techniques for decision-making

PART TWO

Techniques for 'invest or not' decisions

5

5.1 INTRODUCTION

Quantitative techniques or, in general, decision models are a useful aid for arriving at correct decisions and in demonstrating the basis of those decisions to others. The fundamental principle of the techniques is to allow the various relevant factors to be identified, quantified and then combined with the objective of achieving a rational analysis of the problem. Such techniques are also useful to the manager, in that they structure his thought processes and help to ensure that all the relevant factors in each case are considered.

The degree to which decision models for the investment of properties has been developed is limited compared with the progress in other fields. However, the models developed in these fields should be assessed and an appraisal made of the basic techniques used to see whether they can be adapted for use in property investment.

In this chapter we will look into a technique for 'invest or not' decisions – decision analysis or, specifically, the use of decision trees.

5.2 WHAT IS DECISION ANALYSIS?

There are many instances when a decision has to be made without the exact knowledge of every detail relevant to the problem. There are factors associated with the problem which may or may not arise, or whose values may vary over a wide range. The factors may be psychological or rational. The psychological factor relates to the attitude of mind and is subjective. The rational factor consists of those variables that are measurable in one way or another and can be analysed.

Decision analysis pertains to the study of the rational factor and the evolvement of techniques which attempt to clarify and analyse the problem in such a way as to increase the chance of attaining consistent and acceptable results. The essential feature of the analysis is that a decision should be made with an element of chance.

Over the years, new techniques and tools for decision-making have been developed and have assumed sophistication with the increased use of electronic computing. However, techniques cannot minimize the uncertain element of the future, nor guarantee a precise knowledge about outcomes. What can be done is for decision-makers to ensure that the problem is analysed carefully and thoroughly in order to consistently attain the optimal result.

5.3 THE DECISION TREE AS A TOOL FOR DECISION-MAKING

The **decision tree** is one of the tools for decision-making which has remained simple and effective. The method mainly involves the structuring and evaluating of decision problems by presenting the four basic stages of a problem in a logical order, which are:

1. Determining the possible actions or strategies which can be pursued but only one selected.
2. Outlining the events or outcomes which are the consequences of each action as affected by some combination of controllable and uncontrollable factors.
3. Calculating the value or pay-off of the different actions.
4. Choosing the criterion upon which the alternative actions can be judged on the basis of their forecast outcomes.

In doing so, the decision tree is able to incorporate the answers to the questions:

1. What is the objective?
2. What are the alternatives?
3. What is the basis for comparison?
4. What are the possible outcomes of the alternatives?

5.3.1 General

A decision tree presents a decision problem in a diagrammatic tree structure, with four basic components:

1. The *action nodes* – these represent the points in time when a decision-maker needs to select an option from a few alternatives. From each node the different courses of action will take different paths or branches and extend to the right, leading to an event node. Action nodes are usually drawn as squares on the diagram.
2. The *event nodes* – these are the points from which the different possible outcomes are accruing from a course of action. Each outcome will have a branch and lead to a result. These are usually drawn as circles.
3. The *pay-offs* – these are the results of different outcomes to different courses of action. They can be expressed in different units of measure, but for one particular problem all outcomes should be in the same form, so that a comparison can be made.
4. The *probabilities* – these are the likelihood of the future outcome happening. With a series of possible pay-offs, the uncertainties pertaining to each element of the series are of critical importance. The assessment of the uncertainties is by way of probability which ranges from zero to one, representing complete impossibility and absolute certainty. Each outcome is given a probability between zero and one and all outcomes from an event node will have probabilities which add up to one or 100% since one of the outcomes must occur. The assessment of these probabilities can be subjective or objective, depending on how much information of the future is known. The probabilities formulated in a single-stage decision problem are called prior probabilities. The elements are put in a sequential and logical order, from left to right, starting with the initial decision point and ending up with the pay-offs. The base of the tree, which is on the left, represents the time nearest the present, and the tips of the rightmost branches represent the most distant future being considered.

5.3.2 Action vs outcome node

An action even if it is to be taken in the future is under the control of the decision-maker. An outcome can include the reactions of a competitor, the response of a large number of consumers or the state of the economy, the common characteristic being that the outcome is usually a state of nature and is beyond the complete control of the decision-maker. The more an action is under the complete control of the decision-maker, the less uncertainty enters into the decision-making and the outcomes assume greater certainty.

5.3.3 Decision criterion

The selection of the best strategy according to the objective of the decision-maker is based on the criterion which he has laid down. There is no specific criterion which can be regarded as the best because it is determined by the decision-maker's attitude towards risk and uncertainty. However, there are a

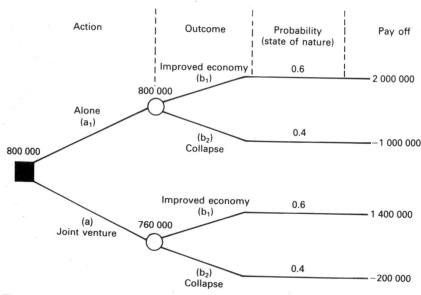

Fig. 5.1 Decision tree for a construction company.

few main criteria which reflect the spectrum of attitudes to risk. These decision criteria are best illustrated using the following example.

A property investment company is considering a contract for a commercial project which will take two years to complete. During the two years, the economy which is on the verge of a recovery can either improve and sustain the growth or face a premature collapse. The company's directors have to decide to do the project entirely alone or enter into a joint venture in order to spread its risk. If the company takes the project alone, it stands to make a profit of £2 000 000 if the economic conditions improved and £1 400 000 if it goes into joint venture. But if the current economic recovery is only a flash in the pan, the company will make a loss of £1 000 000 on its own as against £200 000 in a joint venture (Fig. 5.1).

Let us denote the decision to go ahead alone as a_1 and joint venture as a_2, and the states of nature improved economy and collapse as b_1 and b_2 respectively. The pay-offs for the respective action/outcome combination are presented in the form of a matrix (Table 5.1).

Table 5.1 Pay-off matrix

Outcome / Action	Economy — Improved (b_1)	Collapse (b_2)
Alone (a_1)	2 000 000	−1 000 000
Joint venture (a_2)	1 400 000	− 200 000

5.3.4 Maximin/minimax cost rule

Suppose the decision-maker in the example holds a pessimistic view of life and assumes that whatever action he takes, nature will arrange the worst possible outcome for him. The safest action for him, then, is to consider both the 'Collapse' (b_2) pay-offs and decide on the minimum of the two. Clearly, the action will be to do the project as a joint venture. However, the snag to this rule is that a third action of 'doing nothing' will always become the best action however much the profits of the other outcome because nothing will be lost (Table 5.2).

Table 5.2   Minimax cost

Action \ Outcome	b_2	Minimum loss
a_1	– 1 000 000	
		– 200 000
a_2	– 200 000	
Do nothing	0	0

This rule is therefore doubtful in its application and careful consideration is required.

5.3.5 Maximax rule

This is the other end of the risk spectrum, where the decision-maker always assumes the best outcome whatever the action. Such a decision-maker typifies the 'risk-lover', and in the example he will plum for Action a_1 (Table 5.3)

Table 5.3   Maximax Pay-off

Action \ Outcome	b_1	Maximum
a_1	2 000 000	
		2 000 000
a_2	1 400 000	

5.3.6 EMV approach

Most decision tree analysis is based on the principle of expectation. Given a series of outcomes, each outcome has an assessed probability of occurrence. The analysis will select the action which will maximize the mathematical expectation of all the outcomes. Some outcomes have very high pay-offs (in monetary terms) but little chance of happening, and some have low pay-offs but are likely to happen. The **expected monetary value** (EMV) rule is used to determine by the law of averages which action has the maximum monetary expectation. The two important ingredients are the probabilities assessed and the pay-offs determined for each outcome. The EMV of a particular action is therefore the sum of the pay-offs of all the possible outcomes weighted by their respective probabilities.

In the property investment's decision problem the probability of an improved economy is assessed to be 0.6 and of a falling economy is 0.4, making a total of one (as one of them must occur). Using the pay-offs in Table 5.3, the EMV are:

$$EMV(a_1) = 0.6(2\,000\,000) - 0.4(1\,000\,000) = 800\,000$$

$$EMV(a_2) = 0.6(1\,400\,000) - 0.4(200\,000) = 760\,000$$

The decision, under the above circumstances, should therefore be to do the project on its own as this decision has a higher EMV at £800 000.

The EMV approach is a systematic way of calculating the mathematical maximum expectation which if used correctly will lead to consistent decision-making all the time.

5.3.7 Minimax (regret)/expected opportunity loss approach

An alternative to the EMV approach is to look at the minimization of expected opportunity loss. The regret/opportunity loss for any action/outcome combination is defined as the difference between:

(a) the minimum possible cost (or the maximum possible pay-off) under that outcome; and
(b) the cost (or pay-off) resulting from the action/outcome combination concerned.

Thus the expected opportunity loss of a decision is the sum of all the regrets, weighted by the respective probabilities. In monetary value it represents, in retrospect, the cost–pay-off differential between the course of action which should have been taken and the course of action which was taken.

In the example, the matrix for minimax (regret) is as shown in Table 5.4. The decision is therefore to choose a_2 which has a lower regret. However, the expected opportunity loss which combines the regret (in monetary term) and the probabilities of the outcomes occurring produces the following results:

Table 5.4 Minimax regrets

Action \ Outcome	b_1	b_2	Minimum
a_1	0	800 000	
			600 000
a_2	600 000	0	

$$EOL(a_1) = 0 \times 0.6 + 800\,000 \times 0.4 = 320\,000$$

$$EOL(a_2) = 600\,000 \times 0.6 + 0 \times 0.4 = 360\,000$$

The decision is reversed with the introduction of probabilities. Choosing the criterion of minimum EOL, a decision-maker will opt for a_1 which is the same decision as using EMV rule.

5.3.8 Expected utility value approach

The EMV rule should be adopted for a consistent decision as it will, in the long run, give the best result in monetary terms. However, decision-makers may be influenced by imminent factors pertaining to a particular situation rather than the long-run averages. If the property investment company has limited finances, it may not be able to cope with loss of £1 000 000 but could still hang on with a loss of £200 000, in which case it will probably decide to go into a joint venture so as to minimize its risk within its financial resources.

Another classic example is the case of buying fire insurance policies. Objectively the EMV approach will always show a lower cost for not buying insurance. Psychological factors, such as peace of mind, are not quantified and therefore cannot be built into the EMV calculation. One method for quantifying and measuring a decision-maker's preference pattern for the alternative outcomes arising from the different courses of action open is the use of **utility value**. The utility function (curve) describes an individual's attitudes to risk as represented in Fig. 5.2. The spectrum of attitudes from risk-averse to risk-loving is depicted as functions between utility on the vertical scale and money on the horizontal scale. For the case of the 'risk-averse', there is a diminishing marginal utility for money and vice versa. However, on the whole most people portray different attitudes at different times and a combination of the three curves is representative for all.

The assessment of utility value or utiles for all outcomes of a decision problem is a subjective exercise. Each outcome is given a utility value which reflects a decision-maker's preference and attitude to risk. The utiles are then ranked in order and the optimal decision strategy is to maximize expected utility.

The use of utility as a criterion for decision-making was introduced by Von Neumann and Morganstern in the Standard Gamble. The theory is based on an individual's approach to playing games and how he assesses the outcomes of a game according to utility values. If outcome A is preferred to B which is, in turn, preferred to C, then the expected utility of the gamble of B is:

$$EU(B) = p.U(A) + (1 - p).U(C)$$

where p is the probability of outcome A occurring.

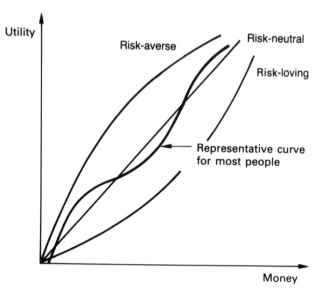

Fig. 5.2 Specimen utility functions.

The assessment of subjective probability to outcomes as in the example can then be achieved if the utility value measuring the pay-offs are 'balanced'. In other words, there is a probability p where the decision-maker is indifferent to the outcome of the Standard Gamble and taking the alternative of a certain outcome B. Another approach to such subjective assessment is to use the 'standard urn' which consists of a certain number of balls of two different colours. Based on the pay-offs, the decision-maker is to decide on the number of white balls out of, say, 100 balls if picking a white ball wins the gamble. The number of white balls is then measured against the number of the different colour balls to obtain the probability which makes the decision-maker indifferent to the outcomes and the stake of the game.

A detailed study of utility function and assessment of subjective probabilities is a subject by itself and is only briefly dealt with here. However, suffice it to say that the criterion of utility value bears a great deal of subjective element. Although there is no fixed rule for assessing utility, each case must be considered consistently in its own circumstances over the period of time.

5.3.9 The roll-back concept

The previous sections have illustrated that, in a decision tree, the decision is made by analysing the pay-offs according to a criterion. The process of moving from the 'top' of the decision tree where the pay-offs are analysed, along the outcome paths where the probabilities are incorporated, and from the outcome node to the decision node along the decision path, is called the **roll-back concept**, or **backward induction**.

In a sequential decision problem the decision tree is rolled back from the furthest point in time to the present. The chronology of events is therefore important and each decision node must be analysed in the proper logical order.

5.3.10 Further examples

Example 1

Consider a contract that calls for bulk excavation work valued at £15 000 to be undertaken during a bad-weather period. The main contractor estimates that given good weather he can carry out the work himself at a cost of £10 000, but if the weather were bad it could cost him £20 000. He knows that a subcontractor will complete the work for £12 000, irrespective of the weather. The contractor has then to decide whether to undertake the work himself or to subcontract.

The decision tree procedure is to write down first the decision point with two alternative courses of action, namely to undertake the work directly or to subcontract the work. If the work is done directly, the cost will depend upon the chance of good or bad weather; in this example, these are put down as simple alternatives. It is necessary to assess probabilities for these alternatives, and this can be based on recorded weather data for the area. In this case, the relative probabilities are taken as 0.7 for good weather and 0.3 for bad weather.

The profit of each possible outcome is calculated and shown in the right-hand column in Fig. 5.3. It is now possible to work backwards from these figures and calculate the expected value of profit for the two alternative decisions:

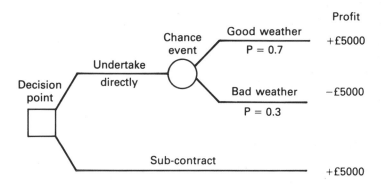

Fig. 5.3 Decision tree.

EMV if work done directly $= 0.7 \times 5000 + 0.3(-5000)$

$$= £2000$$

EMV if work is subcontracted $= 1.0 \times 3000$

$$= £3000$$

On this basis therefore it would appear to be preferable to subcontract the work.

The method can be extended to cover more complex problems involving a series of decisions.

Example 2

An investment company is considering two alternative plans for a high-rise building complex. Plan I calls for a 70-storey apartment building and a separate adjacent 40-storey office building. Plan II envisages the construction of a single, 100-storey building with 45 storeys for offices and 55 storeys for apartments. The estimated costs and life-time returns for the two plans are as follows:

Plan I

Estimated cost, £m.	*Probability*
100	0.6
95	0.3
90	0.1

Estimated life-time return, £m.	*Probability*
300	0.5
250	0.4
200	0.1

Plan II

Estimated Cost, £m.	Probability
150	0.7
120	0.2
100	0.1

Estimated life-time return, £m.	
450	0.2
350	0.4
250	0.3
200	0.1

Which plan should the company adopt? From the decision tree, it would seem to be more profitable to adopt Plan II (Fig. 5.4).

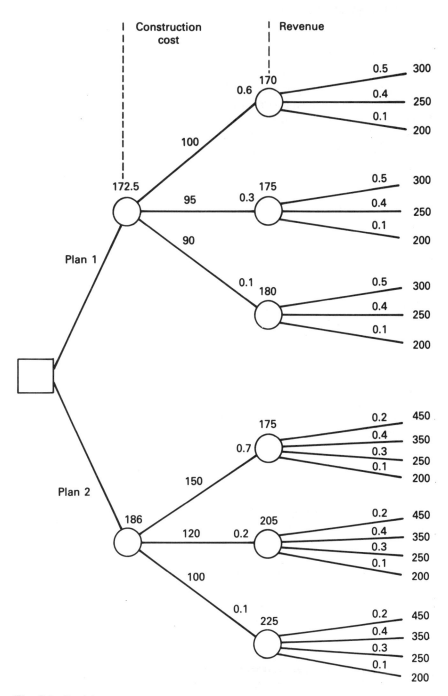

Fig. 5.4 Decision tree.

5.4 THE BASIC AND EXTENSIVE FORM OF A DECISION TREE

5.4.1 The basic structure

The different types of decision problems mentioned earlier are elaborated here. The single-stage, static problem provides a model for the basic structure of a decision tree. The decision is mainly confined to a single problem with one objective, for instance, in the investor's case to decide whether to do it alone or as a joint venture in order to maximize profit. The action node will be confined mainly to the various possible courses of actions representing business strategies, policies and other managerial areas.

The event nodes in the basic structure will show the different chance outcomes to the chosen action. The paths of these outcomes will determine the ultimate pay-offs.

5.4.2 The extensive form

In the real world, decision problems often consist of a series of decisions spread over a period of time before the entire problem is solved. Instances of these multi-stage decision problems abound – a common case being the decision whether to acquire further information in order to attain a more definite picture of how the outcome will be. Using the first example, the contractor is now faced with another decision on whether to conduct a survey to ascertain the level of economic activities in order to predict more accurately the future economic condition. The initial decision is now pushed to the right, and now follows the decision of whether to acquire information. Such a problem is called a sequential and information acquisition decision.

The issue is not so much whether the information acquired will be useful, but whether the cost and time spent in obtaining the information is worthwhile.

In order to find out the maximum cost of information, it is necessary to look at the concept of **expected value of perfect information** (EVPI).

Consider a game consisting of the tossing of a fair coin. The stake of each game is £2. A win will pay £5, and a loss will get nothing. Without any further information, the EMV of playing the game is:

$$(0.5 \times 3) + (0.5 \times -2) = £0.50$$

where the chance of obtaining a head or tail is 0.5, and the EMV for not playing is zero.

Suppose a crystal-ball gazer comes along and offers to predict the outcome of each game for a fee. Assuming that the crystal ball is really able to predict accurately, the EMV with such perfect information would be: EMV (perfect information) = £3 since every time the outcome will be correctly predicted.

The difference of £2.50 represents the maximum amount of the fee that one would be prepared to pay for the information. This is the expected value of perfect information which can be written as:

EVPI = EMVUC (Perfect Information) – initial EMV where EMVUC is EMV under certainty (or with perfect information).

However, perfect information is rarely obtainable but as a theoretical concept helps to set upper limits to the amount that might be worth spending on acquisition of information.

In a decision tree this problem of whether to acquire information can be represented as:

(i) whether to acquire the information or not;
(ii) if information should be acquired, then which source and what level of information to obtain.

In the first case, the problem has one possible initial decision and usually concerns the purchasing of information of a fixed quantity and defined quality at a predetermined price. In the latter case, the decision is usually of a more complex structure decision problem. The event nodes resulting from a decision to acquire information will show the messages from the information source which will, in turn, influence the ultimate outcome.

5.4.3 Sample information

Business decision situations are almost invariably analysed in an environment which is uncertain. To find the true state of affairs precisely additional information may be beneficial to the making of a decision.

Information, however, may be acquired to achieve a partial solution because it is not practical to obtain all information which will result in absolute certainty of the outcomes. The information can be of the form of sample surveys of a market (for product launch), experiments, tests or further research in a particular field. The extra information is then combined with the original or prior probabilities (assessed with existing information) to form revised probabilities or posterior probabilities.

The intention is to use the information to weigh the current assessment or prior probabilities resulting in posterior probabilities, which may still be uncertain but hopefully less uncertain. If presented graphically, the probability distribution function curves should be as shown in Fig. 5.5.

To describe the relationship between the prior and posterior probabilities, the **Bayes theorem** is used.

5.4.4 Bayes theorem

The general form of the theorem is as follows:

If $E_i (i = 1, 2 \ldots r)$ are actually exclusive and the only possible results such that an event F can occur only if one of these r events happens, then the probability that E_j happens when F is known to have occurred is:

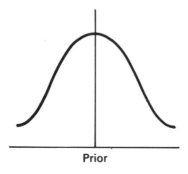
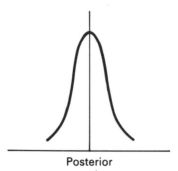

Prior   Posterior

Fig. 5.5 Probability distribution functions.

$$P(E_j/F) = \frac{P(E_j).P(F/E_j)}{\sum\limits_{i=1}^{r} P(E_i).P(F/E_i)}$$

where $P(E_i)$ represents the prior probability of event E;

        $P(F/E_i)$ represents the conditional probability that outcome F occurs, given that event E_i has occurred;

        $P(E_i/F)$ represents the posterior probability of event E_i given that event F has occurred.

5.4.5 Expected value of sample information (EVSI)

One of the aims of calculating posterior probabilities is to help determine the value of information. The **expected value of the sample information** (EVSI) is the difference between the EMV of the optimal decision without information (prior probabilities) and the EMV of the optimal decision with information (posterior probabilities):

     EVSI = EMV (with information) – EMV (without information)

Generally speaking, one would spend £X on obtaining information provided that the difference in the expected monetary value of the optimal decision with information and without information is greater than £X. But the question is how much should £X be?

As in all economic problems, the relative value of information to cost is more of a diminishing relationship than a linear function, with greater proportionate value occurring only at the earlier increments of expenditure. In Fig. 5.6 point Z is the optimum cost/value of information. As cost becomes greater than £X, the acquisition is not worthwhile.

The difference between the EVSI and the cost of information is the **expected net gain from sample information** (ENGSI):

     ENGSI = EVSI – Cost of information

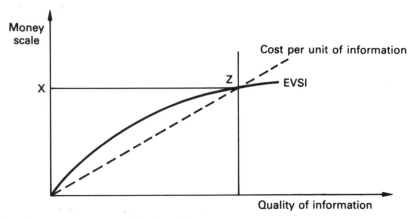

Fig. 5.6 Optimum value of information.

Example:

Returning to the contractor's problem, the directors decided that it may be wise for them to consult a firm of research economists to advise on the level of the current economic activity which will, in turn, indicate the future state of economy. The level of activity is either high or low and the probabilities of continued growth or fall in the economy (outcome) for each level are summarized in the following matrix showing the survey's reliability:

| | | *Current levels of economic activity* | |
		High	*Low*
	Improved (b_1)	0.8	0.2
Economy			
	Collapse (b_2)	0.3	0.7

The matrix of probabilities can also be represented in a diagram shown in Fig. 5.7

Using the Bayes theorem, then the general formula for the calculation of posterior probability is:

$$p(b_1/H) = \frac{p(b_1).P(H/b_1)}{P(b_1).p(H/b_1) + p(b_2).p(H/b_2)}$$

The prior probabilities of the outcome b_1 and b_2 are 0.6 and 0.4 respectively, which are the assessment made without further information. The consultant's report will produce a set of posterior probabilities as a result of the report's indication and the prior probabilities. The revised probabilities (posterior) are then calculated as follows:

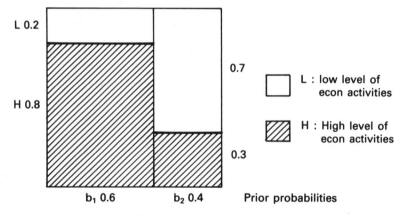

Fig. 5.7 Graphical presentation of probabilities.

$$P(b_1/H) = \frac{0.6 \times 0.8}{(0.6 \times 0.8) + (0.4 \times 0.3)} = 0.8$$

$$P(b_2/L) = \frac{0.4 \times 0.3}{(0.6 \times 0.8) + (0.4 \times 0.3)} = 0.2$$

$$P(b_1/L) = \frac{0.6 \times 0.2}{(0.6 \times 0.2) + (0.4 \times 0.7)} = 0.3$$

$$P(b_2/H) = \frac{0.4 \times 0.7}{(0.6 \times 0.2) + (0.4 \times 0.7)} = 0.7$$

$$P(H) = P(H/b_1) + P(H/b_2)$$

$$= 0.48 + 0.12 = 0.6$$

$$P(L) = 0.4$$

The decision tree for the investor's problem can now be extended as shown in the Fig. 5.8.

The EMV at decision point B is $= 1\ 400\ 000$, whereas at decision point C it is $= 280\ 000$.

The EMVs of B and C are then combined with the probabilities of the consultant's report (0.6 for a 'high' level activity, and 0.4 for 'low' level) to give an EMV of 952 000. The EMV at point D has been calculated and the EMV of 800 000 is posted on it. The decision now rolls back to node A, where the company has to decide whether to hire the consultants. The higher EMV indicates action H without including the consultant's fee. The expected value of sample information (EVSI), which is the optimal EMV (with information) less the optimal EMV (without information) in this example, is therefore £152 000.

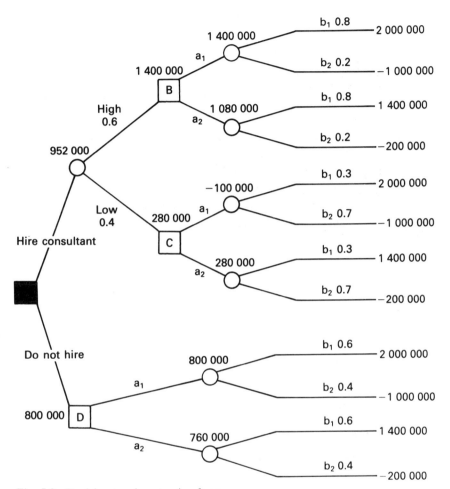

Fig. 5.8 Decision tree in extensive form.

Depending on the cost of hiring the consultant, the directors can then make a decision at node A. If the cost of hiring is less than EVSI, then there is a net gain from the sample information:

ENGSI = EVSI − Cost of information

If the cost of hiring the consultant is more than the EVSI, then it is not worth taking the trouble.

5.5 CONCLUSION

The use of the decision tree in the property field is still to be fully examined, as are most methods of decision analysis. However, with the realization of the need for a more thorough and detailed approach in the application of techniques there is potential for its use in areas, such as property development and investment, which are concerned with the future allocation of resources. The word 'future' in the business world implies risk and uncertainty which are the fundamental elements of decision analysis. To enhance its applicability the decision tree can be incorporated with sensitivity testing in the assessment of subjective probabilities. Further extension may include the use of continuous variables in place of the discrete variables and the application of simulation.

EXERCISE

The parent company of a large corporation is currently looking for a site to build a recreational hall as part of a welfare programme for its staff and the Property Division have been entrusted with this task.

A property consisting of a row of old, three-storey terrace houses has just come on to the market and seems ideal for such a purpose. The Property Division intends to refurbish the ground floor and basement into the recreational centre and the upper floor into staff flats. The likelihood is that planning permission for the change is likely to be granted as surrounding area is undergoing extensive redevelopment.

As the property is in a choice location, it has attracted much interest. The vendor has therefore decided to put the property on the open market for all intending purchasers to submit bids.

The Property Division has assessed the respective chances of winning with a high, medium or low bid as 0.7, 0.5 and 0.2 respectively. If they lose, they would have to purchase an alternative site which would cost £1.5m. in total. The total cost for the choice site would have been as follows according to the bids submitted:

Action	Total cost
Submit high bid	£1.4m.
Submit medium bid	£1.2m.
Submit low bid	£1.0m.

Legal and associated cost involved in the purchase of this choice site is estimated to be £50 000.

The Property Division is meanwhile considering employing a consultant who for a fee of £5 000 would undertake a survey to determine the level of competition for this site – useful information if the Division decides to bid for the site.

The consultant has a reasonable record of reliability but, of course, is not infallible. He will provide a report that indicates whether the competition is strong, moderate or weak and the reliabilities of the outcome according to the report are as follows:

Outcome	Consultant's report			Total
	Strong	Moderate	Weak	
Win	0.1	0.2	0.7	1.0
Lose	0.6	0.3	0.1	1.0

You are to assist the Property Division to arrive at a decision which would minimize total cost.

FURTHER READING

Taffler, R. (1979) *Using Operational Research*, Prentice-Hall, Englewood Cliffs, NJ.

Buchan, J. and Koerigsberg, E. (1963) *Scientific Inventory Management*, Prentice-Hall, Englewood Cliffs, NJ.

Woodward, J.F. (1975) *Quantitative Methods in Construction Management and Design*, Macmillan, London.

Duckworth, W.E. *et al.* (1977) *A Guide to Operational Research*, Chapman and Hall, London.

Thierauf, R.J. and Grosse, R.E. (1970) *Operational Research*, Wiley, New York.

Wilkes, F.M. (1980) *Elements of Operational Research*, McGraw-Hill, New York and London.

Makower, M.S. and Williamson, E. (1975) *Operational Research*, Hodder and Stoughton, London.

Dilmore, G. (1981) *Quantitative Techniques in Real Estate Counselling*, Lexington Books, Lexington, Mass.

Gibson, E.J. (ed.) (1979) *Developments in Building Maintenance: 1*, Applied Science Publishers, London.

6 | Resource allocation

6.1 LINEAR PROGRAMMING

6.1.1 Introduction

There are many situations in the real estate industry where scarce resources must be allocated amongst a number of demands which necessitate optimization. Resources in the context of property development are generally men, machines, materials and money. Mathematicians have developed methods of making allocations with regard to optimizing some criteria which measure the efficiency of allocation. Such methods are generally known as **linear programming**.

In this context, the word 'programming' does not mean setting out a plan of action, as in the case of contract planning or programming, but simply the arithmetical manipulation of numbers, in rather the same way as in computer programming. The word 'linear' simply indicates that the relationship between the factors that are considered can be represented on a graph by a straight line.

In applying linear programming to allocation problems it is necessary to determine an objective. Objectives by which an optimal solution may be selected commonly have a financial basis. One such objective is to make an allocation which minimizes the total cost of carrying out an operation, and another is to allocate on the basis of maximizing the profit. Other objectives could be minimizing total man-hours and distance travelled or maximizing production, storage capacity, etc.

6.1.2 Formulating linear programming problems

Formulating a linear programming problem involves:

(i) identifying the decision variables of the problem;
(ii) expressing each constraint in a linear mathematical form; and
(iii) identifying the objective function.

This procedure is best illustrated by an example.

Example:

A maintenance contractor specializes in the repair and maintenance of two types of flooring. Due to difficulties arising in obtaining raw materials, not more than 280 bays of type X or more than 150 bays of type Y floor can be completed in any one month. The time taken to repair/maintain a standard bay and the contribution to profit of each type of flooring is given below:

Types	Hacking (h)	Relaying (h)	Polishing (h)	Profit (money units)
X	1	10	5	100
Y	2	4	2	200

Due to the men and equipment situation, the maximum number of hours that can be spent on each operation by the firm in any one month is as follows:

	(h)
Hacking of existing floor:	500
Relaying new floor:	3000
Polishing floor:	2250

How should the contractor use his resources to maximize his profit?

6.1.3 Formulation of the problem

The decision variables of the problem are the number of bays of each type of floor to be completed in one month.

Let: X_1 be the number of bays of type X floor; and X_2 be the number of bays of type Y floor

The constraints are the maximum number of hours that can be spent on each operation and the availability of raw materials. These constraints may be expressed as:

Hacking existing floor:	$1X_1 + 2X_2 \leqslant 500$	(1)
Relaying new floor:	$10X_1 + 4X_2 \leqslant 3000$	(2)
Polishing new floor:	$5X_1 + 2X_2 \leqslant 2250$	(3)
Raw materials for type X floor:	$X_1 \leqslant 280$	(4)
Raw materials for type Y floor:	$X_2 \leqslant 150$	(5)

The inequality symbol used in the equations means that an upper limit has been set but no lower limit is required. In addition to the above constraints, the variables X_1 and X_2 cannot be assigned negative values; therefore:

$$X_1, X_2 \geqslant 0 \tag{6, 7}$$

The objective in this problem is to maximize the profit. The objective function is therefore:

$$Z = 100X_1 + 200X_2 \tag{8}$$

Where Z is the maximum profit that can be realized.

Thus the problem has been formulated as a linear programming problem and may be summarized as follows:

Maximize $Z = 100X_1 + 200X_2$

subject to constraints:

$$X_1 + 2X_2 \leqslant 500$$

$$10X_1 + 4X_2 \leqslant 3000$$

$$5X_1 + 2X_2 \leqslant 2250$$

$$X_1 \leqslant 280$$

$$X_2 \leqslant 150$$

6.1.4 Graphical solution of linear programming problems

Linear programming problems expressed in two variables can readily be solved using graphical methods. Problems involving three or more variables, however, cannot be solved in this manner, and for such problems a matrix solution is used.

By considering the seven constraints in the example considered earlier, the feasible solution for X_1 and X_2 could be any point which lies in the region shown shaded in Fig. 6.1. The region is identified by plotting the five inequalities as equalities on a $x_1 - x_2$ plot, and the inequalities help to determine which side of the equations the feasible region will lie.

The solution to the problem requires the objective function $Z = 100X_1 + 200X_2$ to be maximized. If this equation is plotted with different values for Z, they will all be parallel straight lines. The maximum value of Z which also satisfies the constraints is the optimum value. This is achieved by visualizing the movement of the line as Z is increased. A line which crosses the feasible region will cut the corners at two points, and the value of Z increases as the line moves upwards (to the right). It can be concluded that the maximum value of Z is when the line passes through one (or more) of the corners of the feasible region.

In this example, the objective function attains the maximum value when it passes through B or C or any other point on the line BC in Fig. 6.1.

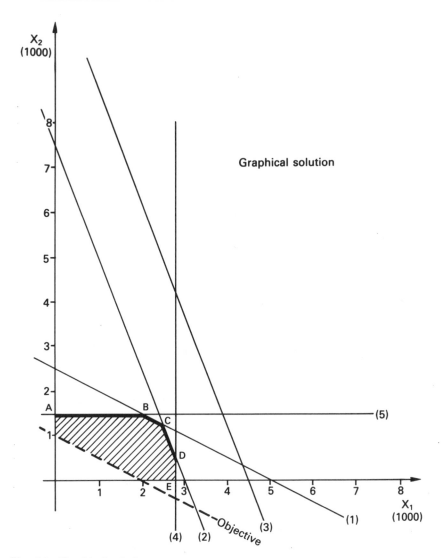

Fig. 6.1 Graphical solution.

The coordinates of B and C can be determined by solving simultaneously the equations which intersect at B and C; from the graph in the figure these are:

B: $X_1 = 200 X_2 = 150$

C: $X_1 = 250 X_2 = 125$

The profits that can be made are:

at B = 50 000

at C = 50 000

Therefore at either B or C the contractor is maximizing his profits.

6.2 SIMPLEX TECHNIQUE

The graphical technique dealt with earlier cannot handle more than two variables unless it is extended into a three-dimensional visualization. An alternative solution of the problem is the **simplex method** which requires a matrix
manipulation in obtaining an optimum solution. Any number of variables can
be handled in this method. The explanation of this technique is somewhat
cumbersome and could lead to confusion. It is sufficient to state here that many
of the computerized models of linear programming are based on this technique.

6.2.1 Minimizing

It is often necessary to minimize a function such as cost. Minimization of an
objective function can be achieved by transforming the

$$Z = X_1 + X_2 \text{ into } (-Z) = -X_1 - X_2$$

Consequently, a solution can be obtained in exactly the same manner as
previously.

6.2.2 Assumptions

There is no workable management technique that takes nothing for granted.
Every device that is used from advanced mathematical programming or statistics at one extreme, to business experience and raw hunch at the other, will
make assumptions. A model is supposed to simplify reality. If it does not
simplify, it is not a model. However, when using a technique it is wise to be
aware of the main assumptions that are made. In the case of linear programming these are as follows.

(1) All data known and constant

This means that there are no unknowns, no random variables and no dynamic
variables (values changing with time). It means that prices of outputs and
factors are constant and unaffected by output or usage. It means that the firm
knows its objective and can measure it. There can be no interdependencies
between prices themselves or prices and outputs. The firm operates in a 'fixed

price' situation, but the specific, strong assumption of 'perfect competition' in factor and product markets is not made.

It further means that technical coefficients are fixed – the amounts of resources needed per unit of each X are given. Thus economies or diseconomies of scale are not allowed in the basic model. Finally, the assumption means that management has predetermined the possible range of activities.

(2) No fixed charges

A 'fixed charge' is a set-up cost; it is only incurred if an activity is used. For instance, to start a production run of a product machinery may have to be re-set. The re-set costs are not incurred if the product is not made.

(3) Divisibility

It is assumed that fractional values of the Xs are meaningful.

(4) Boundedness

There must be linear restrictions on the choice of values of the Xs such that the objective function cannot become infinite. In practice, these restrictions take the form of linear inequality constraints, each involving some or all of the Xs and sign requirements. In fact sign requirements are not necessary mathematically, but the Xs must be bounded from below and from above.

Summary

It is often said that linear programming is possibly the single most important quantitative management technique. It has been of great value in manufacturing industry particularly; many computer companies offer a range of linear programming packages suitable for a wide range of problems from shopfloor production to overall corporate modelling. Many of the non-linearities of the real world can be satisfactorily approximated by linear functions. Sensitivity analysis is most important and is a practical way of taking uncertainties into account. Any problem which has the character of maximization, subject to constraints, is a programming problem of some sort. All significant economic problems involve constraints and, in most cases, best rather than merely adequate arrangements are sought. In property investment the efficient use of scarce resources should also be carried out by the application of linear programming.

EXERCISES

There are two problems which typify the use of linear programming in building and estate management work. The reader is now invited to try to solve them.

Exercise 1

A company is engaged in the business of supplying gravel and sand to various construction projects. The company owns two quarries. Analysis shows that material from each quarry has the following composition:

Quarry	Coarse gravel %	Fine gravel %	Sand %	Waste %
A	20	20	30	30
B	30	30	30	10

Material from quarry A costs £8 per cubic metre for extraction, hauling and separation into various components, whilst that for quarry B costs £10 per cubic metre. The prevailing market prices per cubic metre for coarse gravel, fine gravel and sand are £15, £15 and £20 respectively.

The management estimates that the potential market during the next month will be as follows:

For coarse gravel *not less than* 25 000 m^3
and for fine grave *not more than* 36 000 m^3

The plant for separation of the material into various components has a capacity of 150 000 m^3 per month. Determine the company's optimal plan for material extraction for the next month.

Exercise 2

A developer wants to build a condominium comprising two-bedroom studio flats (high-rise), three-bedroom apartments (medium-rise) and four-bedroom maisonettes (low-rise). The total number of units he can build is 100, and to ensure a proper mix of house types the planning authority has stipulated that there should be no more than 30% of high-rise units.

From his building consultant, he estimated that he can build 30 units of studio flats, 20 units of apartments and 15 units of maisonettes in one year. In order to make a profit, the developer must finish the project in two years.

Due to the greater demand for smaller units and their lower land cost per unit, the developer estimated that his profits from each type are as follows:

Profit per unit

2 bedroom	£7000
3 bedroom	£6500
4 bedroom	£6000

The developer wants to maximize the profit from this venture. Set up the problem into the appropriate mathematical form to solve the problem using a linear programming technique.

6.3 THE TRANSPORTATION PROBLEM

6.3.1 Introduction

The transportation technique is a method to tackle a group of problems which may also be solved by other linear programming methods. The technique has been derived and is largely used by distribution industries. It is a purpose-built algorithm that is similar to the simplex method.

The method is typified by the problem situation where a company has goods in a number of depots around the country and has to distribute them to customers in various locations. In this case, the constraints are that at each depot there is available no more than a certain amount; and an exact amount has to be delivered to each customer. The objective function here is to minimize the total transport costs. An important point to note in transportation problems is the absence of any economies or diseconomies of scale.

The delivery of materials, such as concrete aggregates, cement and bricks from various sources to different sites, are examples of this problem which may be tackled by the transportation method.

Example:

A property management firm has five service depots to cater to the maintenance needs of four clients. Each client has an average number of work orders per month which have to be satisfied by the service depots. However, because of resource constraints, each depot can only have a limited service capacity – i.e. the number of work orders it can satisfy in a month. The costs and quantities are given in tabular form below:

Client	*Service depot*					*Average no. of work orders per month*
	P	Q	R	S	T	
A	5	10	9	7	5	20
B	8	11	7	9	12	25
C	10	6	7	5	9	10
D	7	9	8	10	11	16
Service capacity	12	15	22	9	13	

Determine the optimal allocation to minimize cost.

Solution:

This is referred to as a balanced problem since the total supply and total demand are equal. The method to be adopted, amongst various other algorithms, is known as the **Vogel's Approximation Method** (VAM), which follows the logic of the solution procedure in the simplex method. Briefly, the first step is to establish a basic feasible solution, followed by checking for improvement possibility and, thirdly, implementing the improvement and continuing until no further improvements are possible.

Step 1   Determine the VAM number for each row and column of the tableau, which is the difference between the two most economical routes in that row or column (Fig. 6.2).

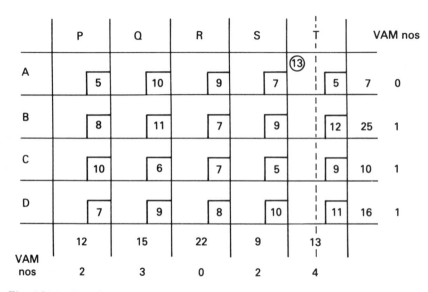

Fig. 6.2(a)   Vogel's Approximation Method.

The initial allocation is determined by the largest VAM No. and dispatch as much as possible along the cheapest route in the row or column selected. The maximum assignment will be given by the smaller of the supply and demand figures corresponding to the cheapest square, which in the example is col. 5, row 1, and 13 is entered into the square.

This fact is indicated by broken lines in that column – i.e. Service Depot T has exhausted its capacity (but Client A still needs 7). A second set of VAM Nos. has to be calculated ignoring the deleted column (or row), which in our case will mean column 5.

The second set of VAM Nos has col. 2 giving the highest number (i.e. 3) and following the same procedure, square C, Q is entered with the figure 10, indicating Client C is now satisfied and Depot Q still has 5.

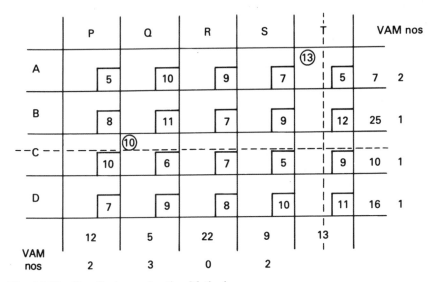

Fig. 6.2(b) Vogel's Approximation Method.

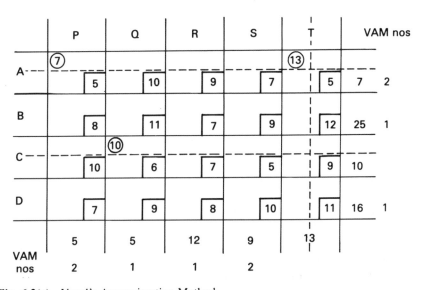

Fig. 6.2(c) Vogel's Approximation Method.

The procedure is repeated until all the allocations have been carried out. In case there is a tie in the VAM No., it is up to the user to decide (but normally one would choose the least-cost square). In the example, there is a three-way tie in the next repetition and since the square A, P has the least cost, the remainder 7 orders from Client A is now satisfied.

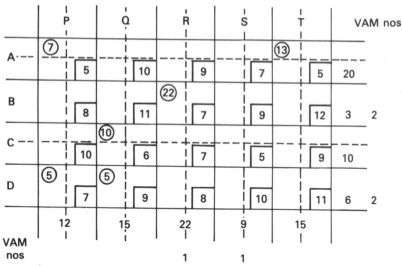

Fig. 6.2(d) Vogel's Approximation Method.

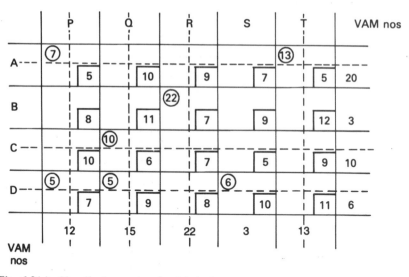

Fig. 6.2(e) Vogel's Approximation Method.

As can be seen, the procedure is repeated eight times before we arrive at the VAM initial solution. The total cost of this arrangement will be:

$$\text{Cost} = (7 \times 5) + (13 \times 5) + (22 \times 7) + (3 \times 9) + (10 \times 6) +$$
$$(5 \times 7) + (5 \times 9) + (6 \times 10)$$
$$= £481$$

	$V_1 = 5$ P	$V_2 = 7$ Q	$V_3 = 6$ R	$V_4 = 8$ S	$V_5 = 5$ T	
$U_1 = 0$ A	⑦ 5	10	9	7	⑬ 5	20
$U_2 = 1$ B	8	11	㉒ 7	③ 9	12	25
$U_3 = -1$ C	10	⑩ 6	7	5	9	10
$U_4 = 2$ D	⑤ 7	⑤ 9	8	⑥ 10	11	16
	12	15	22	9	13	

Fig. 6.3 Vogel's Approximation Method.

Step 2 The question now is whether or not the total cost of £481 can be reduced (Fig. 6.3).

From the solution tableau (Fig. 6.3), it can be seen that there are squares which actually have lesser costs but are not utilized. The first thing to do in the procedure to improve the initial solution is to work out the U, V values for each of the rows and columns. The U values are for the rows, and V values for the columns. Insert the A row with a U value of zero and determine the rest by the costs of the assigned squares such that $U + V = $ cost. In the example, $V = $ Cost of AP square $- U$ or $V_1 = 5$ and $V_5 = 5$. If $V_1 = 5$, then $U_4 = 2$. If $U_4 = 2$, then $V_2 = 7, V_4 = 8$. If $V_2 = 7$, then $U_3 = -1$. And if $V_4 = 8$, then $U_2 = 1$ and subsequently $V_3 = 6$.

Once the U, V values have been determined, we can calculate the $(U + V - \text{Cost})$ values for the squares which are not assigned.

These values are entered in Fig. 6.4. If all these values are negative or zero, then the last allocation matrix is optimal. But in this case there are two squares with non-zero positive values. This exercise of calculating U, V values is actually to determine the opportunity costs (next minimum cost), so that the allocation will be more efficient. The U, V values are sometimes known as 'dummy costs'. The evaluation of the dummy costs of the non-assigned routes will show if there could be a further gain (reduction in cost) or improvement to the initial solution. Negative or zero values will indicate that there is no more improvement.

The next step is to assign the square which shows the largest positive value which in the example is cell CS. The logic should then follow that if some

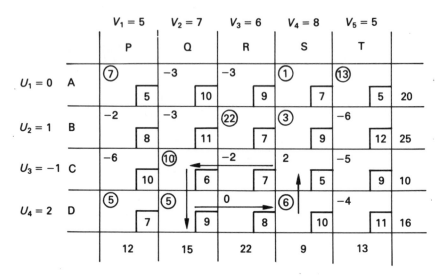

Fig. 6.4 Vogel's Approximation Method.

assignments are to be made from S to C some of those 'existing' routes must reduce by the amount. To decide which of the currently assigned squares should reduce, it is best to use a 'closed path' (or modified distribution method). The rules to follow are that the path must start and close at the selected square (CS); it must go in one direction and it can turn only at an assigned square. What these mean is that the redistribution will in fact still satisfy the requirements (supply and demand). The path to be taken in the example should thus follow the arrow in Fig. 6.4.

Let us assign the 'new amount' to be x. If x is to be assigned at CS, then following the arrow, square CQ must reduce by x since Client C only needs 10 work orders. If CQ is reduced by x, then DQ must increase by x, based on the same argument, and DS must reduce by x. But what should x be? It should be the smallest assigned value of all the squares to be reduced by x. In this case, it is 6.

The new solution should then be as shown in Fig. 6.5. The new cost will then be:

$$\text{Cost} = (7 \times 5) + (13 \times 5) + (22 \times 7) + (3 \times 9) + (4 \times 6) +$$

$$(6 \times 5) + (5 \times 7) + (11 \times 9)$$

$$= \pounds 469$$

which shows that there is an improvement.

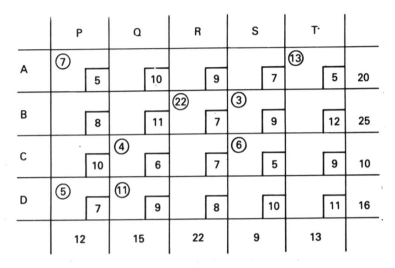

	P	Q	R	S	T	
A	⑦ 5	10	9	7	⑬ 5	20
B	8	11	㉒ 7	③ 9	12	25
C	10	④ 6	7	⑥ 5	9	10
D	⑤ 7	⑪ 9	8	10	11	16
	12	15	22	9	13	

Fig. 6.5 Vogel's Approximation Method.

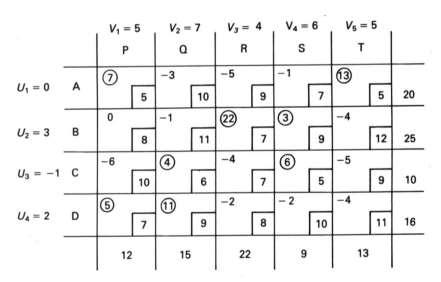

		$V_1 = 5$ P	$V_2 = 7$ Q	$V_3 = 4$ R	$V_4 = 6$ S	$V_5 = 5$ T	
$U_1 = 0$	A	⑦ 5	−3, 10	−5, 9	−1, 7	⑬ 5	20
$U_2 = 3$	B	0, 8	−1, 11	㉒ 7	③ 9	−4, 12	25
$U_3 = -1$	C	−6, 10	④ 6	−4, 7	⑥ 5	−5, 9	10
$U_4 = 2$	D	⑤ 7	⑪ 9	−2, 8	−2, 10	−4, 11	16
		12	15	22	9	13	

Fig. 6.6 Vogel's Approximation Method.

But can it be improved further? To answer this question, we must return to Step 2 and repeat the whole procedure (Fig. 6.6).

From Fig. 6.6, it can be seen that no $(U + V - \text{Cost})$ value is positive and therefore we can conclude that the last solution is the optimal solution and that £469 is the least cost.

6.3.2 Other issues of transportation problems

- Often the objective is to maximize rather than minimize a function. In such cases, basically the same procedure is adopted, except when selecting cells of smallest cost it would have to be cells with greatest profits, i.e. a reversal of values.
- Where there is unequal supply and demand –
    - (a) if total supply exceeds total demand, an additional dummy row must be added to the matrix with zero costs of transport;
    - (b) if total demand exceeds total supply, an additional dummy column is added up to the matrix in order to satisfy this excess demand, the transport costs for the cells in this column being zero.

Resource planning | 7

7.1 The assignment problem
7.2 Resource planning: scheduling
7.3 Resource planning: sequencing

7.1 THE ASSIGNMENT PROBLEM

7.1.1 Introduction

The assignment problem is a special case of transportation problem in which there is only one item at each dispatch point and only one item required at each receiving point. This means that there must be equal numbers of dispatch and receiving points and also routes.

The method is used for problems such as:

(a) The assignment of n gangs of men to m tasks that they carry out in the minimum of time.
(b) The assignment of n project managers to m projects, so that the probable overall profit is maximized.
(c) The assignment of n delivery trucks to m delivery routes, so that overall distance covered is minimized.

The method is best illustrated by an example.

Example:

Taking a simplified case, as in (a), above, the following matrix shows the time taken by each man (A, B, C, D and E) to do each job (a, b, c, d and e). The objective is to find out who does which job, so that the total time is at a minimum.

	A	B	C	D	E
a	3	5	10	15	8
b	4	7	15	18	8
c	8	12	20	20	12
d	5	5	8	10	6
e	10	10	15	25	10

Solution:

The first step is to deduct the smallest element in each row from all elements in the row as in the following matrix:

	A	B	C	D	E
a	0	2	7	12	5
b	0	3	11	14	4
c	0	4	12	12	4
d	0	0	3	5	1
e	0	0	5	15	0

The second step is to deduct the smallest element in each column from all elements in the column as in:

	A	B	C	D	E
a	0	2	4	7	5
b	0	3	8	9	4
c	0	4	9	7	4
d	0	0	0	0	1
e	0	0	2	10	0

The next step is to make a 'zero assignment' by drawing lines through the elements of the matrix in such a way to cross out all zeros with the *minimum* possible number of lines. In the above matrix only three lines are required to cross all zeros, and since this is less than the number of rows, a zero assignment has not been achieved. The next step is to find the smallest uncrossed element and deduct its value from all uncrossed elements and add its value to all elements cut by two lines (one vertical and one horizontal). This means all the uncrossed elements will be reduced by 2, whilst the ones crossed by two lines increased by 2, as shown here:

	A	B	C	D	E
a	0	0	2	5	3
b	0	1	6	7	2
c	0	2	7	5	2
d	2	0	0	0	1
e	2	0	2	10	0

Another attempt is made to draw lines through the matrix to make a zero assignment, this time resulting in four lines. The procedure is repeated in the two following matrices until the minimum number of lines required to cross out all zero elements equals the number of rows (or columns):

	A	B	C	D	E
a	1	0	2	5	3
b	0	0	5	6	1
c	0	1	6	4	1
d	3	0	0	0	1
e	3	0	2	10	1

This time it has been necessary to use five lines to cross all zeros and therefore a zero assignment has been achieved.

	A	B	C	D	E
a	1	0	0	3	3
b	0	0	3	4	1
c	0	1	4	2	1
d	5	2	0	0	3
e	5	0	0	8	0

The actual assignment is determined as follows: column 5 has only one zero and therefore E will be assigned to job e. Other assignments to E should be ignored. Similarly, the other assignments $D - d$, $C - a$, $B - b$ and $A - c$ can be determined easily.

The total distance travelled is:

$$(C - a) + (B - b) + (A - c) + (D - d) + (E - e)$$

$$= 10 + 7 + 8 + 10 + 10$$

$$= 45 \text{ units}$$

7.1.2 Maximization

If the matrix element represents an objective which is to be maximized, for example, profit, then such a matrix can be converted for treatment as a maximization problem by subtracting each of the elements from the largest element in the matrix. A modified matrix of differences will then be obtained.

Non-square matrix problems can also be dealt with by adding dummy columns or rows with zero elements.

7.1.3 Summary

The use of transportation technique is quite extensive in solving management problems. For property investments, it may be applied to the area of financing.

However, financing problems will typically have the structure of more complicated programming problems and, ideally, the problems of optimal sources and uses of funds should be examined simultaneously. This will not always be possible, but occasionally computational techniques originally designed for quite different management applications can suit the structure of simplified financing problems.

There are numerous changes that can be rung on the basic model. In transshipment problems there are intermediate locations through which goods may pass. In capacitated problems there are limits on the amounts that may be sent along some or all routes. Diseconomies of scale can be allowed for (a discrete increase in unit cost along a route if used above a certain level). Unfortunately, economies of scale, which are perhaps more frequent, are much more troublesome computationally. Nevertheless, the range of uses of the method is considerable and also that the basic model is highly adaptable.

EXERCISES

Exercise 1

A civil engineering contractor is engaged to carry out earthmoving work for the construction of a new section of an expressway. Fill material can be supplied from three borrow pits P, Q and R located near the works up to a maximum of 60 000 m^3, 80 000 m^3 and 100 000 m^3 from each respectively. The material is to be delivered to three locations A, B and C along the road and each requires 90 000 m^3, 50 000 m^3 and 40 000 m^3 respectively. The costs in pounds per cubic metre for delivery of the material from the pits to each of the three sites is shown below; determine the best arrangements for supply and delivery of the material, if the objective is to minimize cost:

		Pits		
		P	Q	R
	A	7	5	6
Sites	B	4	3	5
	C	10	9	9

Exercise 2

A contractor has been successful in obtaining five new projects. The projects, however, are different in value, type of work and complexity. As a result, the experience and qualities required of the project manager for each will be different. After careful consideration, five managers are selected and their skills assessed against each project. Each manager is scored on a points scale with a maximum of 100 marks indicating that the manager is highly suitable, and a zero mark that he is unsuitable for the work. The individual assessments are shown below:

		Manager			
	1	2	3	4	5
A	75	28	61	48	59
B	78	71	51	35	19
Project C	73	61	40	49	68
D	55	50	52	48	63
E	71	60	61	74	70

Which managers should be allocated to which projects if the company wishes to distribute them in the most effective way?

7.2 RESOURCE PLANNING: SCHEDULING

7.2.1 Introduction

One of the most important objectives of property investment has been the proper use and allocation of resources. In a world of limited resources, investors have always sought to optimize with whatever they have, men, materials, money and machines, in order to meet the eventual target of maximum profits or minimum cost.

7.2.2 Resource smoothing/resource levelling

In the first instance, let us assume that there are unlimited resources. In such a situation, only the sequence of the activities matters – i.e. the interdependencies of the operations which may entail certain operations to start only after the completion of some specific operations. Concurrent activities will have no restriction and can be carried out at the same time. The activities/operations can therefore all start at their earliest start time, or latest start time.

If all the activities start at their earliest possible time, then there will be a concentration of the use of resources during the early stages of the project, as shown in Fig. 7.1. If all the activities start at the latest possible time, the concentration will then be on the later stages of the project.

To avoid such concentration of the use of resources, which may cause unforeseen problems to management, the resources should ideally be smoothed out or evenly spread over the project duration. This distribution of resources is **resource smoothing**. For most construction/maintenance operations, perfect resource smoothing is impossible. In fact, for most operations, the ideal situation of using the same amount of resources throughout the project is difficult to achieve.

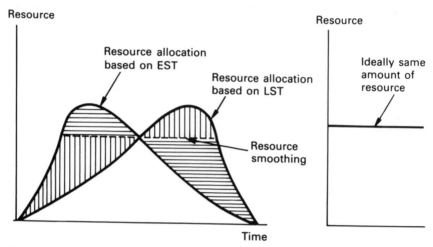

Fig. 7.1 Resource smoothing.

However, even where operations are amenable to resource smoothing, the resources themselves are usually in limited supply. That being the case, the technique of scheduling is more appropriately considered as **resource levelling** rather than resource smoothing.

In Fig. 7.1, if the maximum limit of resources is below the resource smoothing line, it is obvious that the total project duration would inevitably increase. On the other hand, if the maximum limit is above the resource smoothing line, then the limit is more than what is required.

Scheduling of operations/activities is therefore imperative to resource levelling. Under conditions of resource constraints, the activities will need to be scheduled carefully, so that the total duration is at its shortest.

7.2.3 Methods of solution

Resource allocation, or specifically the scheduling of operations, has been widely researched in the building industry. The main methods of solution available can be classified under four main groups.

(1) Analogue methods
Analogue methods rely on the ability of the human brain to solve combinatorial problems, and do no more than provide it with suitable aids, that is analogues of the problem on which the brain can work.

(2) Heuristic methods
Heuristic methods use empirical decision rules which produce a better-than-average allocation, but not necessarily the best. The rules can be set out in

formal terms and therefore programmed on a computer. These methods are the most widely applied in practice.

(3) Mathematical programming methods

Mathematical programming sets out all the restrictions and relationships in the problem as a number of simultaneous equations. From the infinite number of possible solutions to these, it selects the one which optimizes the objective function (Chapter 6).

(4) Profile functional methods

Profile functional methods study the overall shapes of resource profiles and describe them as mathematical functions. The parameters of the functions may then be adjusted as a means of allocating total resources.

The two main methods which we will look into are the serial and parallel methods. These methods can be classified as heuristic because they are basically resource levelling procedures or algorithms.

(5) Serial and parallel methods

Before we explain the use of these methods by examples, we shall examine some of the decision rules that can be adopted by various heuristic methods. Whenever the required amount of a resource exceeds the amount available, a choice must be made between the jobs which are competing for the limited resources. The choice is made by a decision rule (or rules) which applies an index of priority to each job. The jobs are ranked or sorted in order of priority, and the resources are then allocated in this order until they are exhausted.

All decision rules in practical use include at least one of the following sorting criteria:

- *Total float*: because the most critical jobs should be done first.
- *Earliest start*: 'Get on with the job just as soon as you can' – this may not be a very useful measure of priority because there is often a flood of initial jobs demanding resources, all having the same earliest start.
- *Latest start*: this often turns out to be the same criterion as total float.
- *Earliest finish*: this is related to 'earliest start', being the same as the earliest start of the immediate successors of the job under consideration.
- *Latest finish*: the same as the latest start of the immediate successors.
- *Duration*: doing the shortest jobs first will give the earliest opportunity for revising a decision.

In the **serial method of resource allocation**, the general principle is that the activities are sorted into a list, so that any activity in the list has those preceding it placed above it, and those succeeding it below it. The principle of allocation is such that one activity is dealt with at a time, working in sequence down the list. As the activity is considered, an allocation of resource is made to it, and so on until each activity in the list has received its respective allocation. From the method of working it will be seen that the method can easily be used for more than one project, using common resources simply by listing all the activities properly sorted.

In the **parallel method**, each period of time is considered in turn rather than each activity. The available resources in any one time period are allocated on the basis of some criteria which set up priorities. For example, it may be that, if two or three or more activities compete for the available resources, the resource will be allocated to that activity with the least total float, the total float being used as a measure of the criticality of any activity. Having allocated the resources which are available for a particular time period, the allocation moves forward to the next time period and the activities which have not previously received an allocation of resource are then considered.

Due to the fact that consideration is made on one time period later in the programme, the criticality of those activities, which have already been considered but rejected in the light of other more critical competition for the resources, has increased. For example, if an activity cannot obtain an allocation of resource that has a total float of ten weeks at one particular time period, then in considering the next time period of one week the total float of the activity brought forward will be nine weeks, and so on until it is in such a position that its total float enables it to obtain priority over all other activities which are brought forward.

In such a system, a situation may arise where three critical activities are competing for the resources which are sufficient only to satisfy two of them. In this case, one of the critical activities will have to be put back for consideration in the next time period and inevitably, since the activity is critical, the overall duration of the project will be extended.

EXAMPLES

Unlimited resources

Figure 7.2 shows a network diagram for a small maintenance operation. In the case of unlimited resources, the decision rule for resource allocation can be any of the six mentioned. Suppose we adopt the earliest start time for each activity, we will then proceed to allocate the resources to all the activities that can start, following the network logic or the dependency table. The resource histogram for the allocation is shown in Fig. 7.3. The project can be completed in 51 days (as in the network) using a maximum of 14 units of resources.

Limited resources – serial method

The serial method of resource levelling, in this example, is based on the condition of latest finishing time (LFT). The first step in the procedure is therefore to work out the LFTs of every activity and rank each activity according to its LFT. The

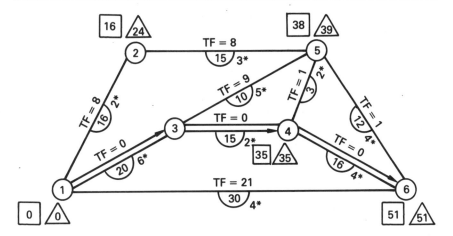

n* = No of resource units required
TF = Total float

Fig. 7.2 Network diagram for a small maintenance operation.

ranking is done in ascending order of the LFT – i.e. the smaller the LFT, the smaller the ranking. In case of equal LFTs, break the tie by comparing the total float of the activities and, again, the smaller ranking goes to the smaller TF.

The next step is then to allocate each activity using a histogram until the maximum resource level is reached (in the example, let's say the resource limit is 10 units). The allocation is done according to the ranking and should follow the logic of the network. An activity cannot be split in time – i.e. it must be carried out continuously. The histogram (see Fig. 7.3) shows that the project duration is now expanded to 80 days (Table 7.1).

Table 7.1

Activity	Duration	LFT	TF	Ranking	Required resources
1–2	16	24	8	2	2
1–3	20	20	0	1	6
1–6	30	51	21	9	4
2–5	15	39	8	5	3
3–4	15	35	0	3	2
3–5	10	39	9	6	5
4–5	3	39	1	4	2
4–6	16	51	0	7	4
5–6	12	51	1	8	4

Limited resources – parallel method

For the parallel method, again, we have chosen the LFT as the criterion for the allocation of resources and the activities are ranked in ascending order of the LFTs (Table 7.1). In case of a tie, the TF of the activities are compared and a smaller ranking is given to the smaller TF.

The next step involves the setting out of a table for the allocation procedure. In Table 7.2 the heading 'trigger event' means the event which will lead to the start of activities (as according to the network logic). The 'time to finish' column is computed by adding the activity duration to the time shown in brackets under the trigger event; the rest of the columns are self-explanatory.

Table 7.2

Iteration no.	Trigger event	Activities can start	LFT	Rank	Duration	Time to finish	Resources required	Cumulative resources employed
1	Start (0)	1–3 —	20	1	20	20	6	6
		1–2 —	24	2	16	16	2	8
		1–6	51	9	30			
2	Finish of 1–2 (16)	2–5 —	39	5	15	31	3	9
		1–6	51	9	30		4	
3	Finish of 1–3 (20)	3–4 —	35	3	15	35	2	5
		3–5 —	39	6	10	30	5	10
		1–6	51	9	30		4	
4	Finish of 3–5 (30)	1–6 —	51	9	30	60	4	9
5	Finish of 2–5 (31)							
6	Finish of 3–4 (35)	4–5 —	39	4	3	38	2	6
		4–6 —	51	7	16	51	4	10
7	Finish of 4–5 (38)	5–6	51	8	12		4	
8	Finish of 4–6 (51)	5–6 —	51	8	12	63	4	8

Note:
Subject to a maximum of 10 units; broken lines indicate that activities are allocated with resources; total project duration: 63 days (or the latest finish time).

Some of the rules for the allocation are:

(a) Allocate in logical sequence of trigger event and in accordance with ranking.

(b) Within one iteration (determined by the trigger event) an activity cannot start, even if there is sufficient resources, unless the activities which are ranked before it have started. In other words, if an activity which has a prior ranking cannot start due to insufficient resources (as dictated by the cumulative resource employed), then proceed to the next iteration, even if the subsequently ranked activities within the same iteration can start because they require less resources.

(c) Each iteration is marked by the finishing of one activity and thereby releasing resources.

(d) An activity must be continuously carried out and not split in time.

The allocation using the parallel method is shown in Table 7.2. Subjecting to a maximum limit of 10 units of resources, the project can be completed in 63 days. In fact the progress of the project can be gauged by the LFTs of each job and the time to finish. The project is delayed by 9 days at iteration no. 4 and subsequently delayed again in the last iteration by another 3 days.

7.3 RESOURCE PLANNING: SEQUENCING

7.3.1 Objectives of sequencing

As the term implies, 'sequencing' has to do with the order in which jobs are carried out. Sequencing-type problems – also known by the fuller, more descriptive title of 'job shop sequencing' problems – originated in a factory shopfloor situation, where the problem is that of determining the order in which several jobs shall be done on several machines.

However, the problems are not confined to the factory production-line setting. For instance, a building firm may have contracts at a number of sites, all of which call for the use of excavators, cement mixers and such like equipments – of which there are only a limited number. How should the contracts be worked on to minimize completion time or to secure good equipment utilization?

Some of the objectives of sequencing problems are listed below:

- Minimizing the total time jobs are waiting to start using the machine/equipment.
- Minimizing total time between starting the first job and completing the last (or minimizing total elapse time or makespan).
- Minimizing total lateness (or tardiness), which is the sum of all lateness in each job (the jobs having deadlines to meet).
- Minimizing maximum lateness or tardiness.
- Minimizing the number of jobs which are being late.
- Minimizing the cost of being late (i.e. penalties to be paid for lateness).
- Minimizing the storage cost whilst carrying out the project.

Workable approaches to this problem involve priority rules – e.g. process that job first with the shortest processing time, or with the earliest late start time. These are empirical rules which have been tested and compared in simulations and will suit certain objectives. But there is no workable technique that provides a global optimum.

Obviously, some of the objectives can be solved readily by a commonsense judgement or rule-of-thumb. Some problems are more complicated, for instance, a problem in which five jobs must be processed on each of six machines. In the absence of precedence requirements, there are (6!) alternative solutions (or 2 985 984 000 000). Such problems are, of course, combinatorial in the extreme.

Some of the objectives also seem to be duplicative, for example, minimizing maximum lateness, minimizing number of jobs which are being late and minimizing total lateness. The following matrix should clear up the differences.

Table 7.3   Lateness of five jobs in three different projects

	A	B	C	D	E	Total lateness	Maximum lateness	No. of jobs being late
Project 1	3	21	2	7	9	42	21	5
Project 2	7	8	15	15	18	63	18	5
Project 3	Early	35	Early	10	Early	45	35	2

In the following sections, we explore the priority rules (or solutions) which will satisfy some of the objectives.

7.3.2 Minimization of total waiting time in the case of one machine and multiple jobs

The solution to this situation is a simple one. Arrange the jobs in order of processing time from the shortest to the longest. This is also the order for the jobs, so that the total waiting time is minimized.

7.3.3 Minimization of total elapse time

In the case of two machines, where the use of machine A must precede the use of machine B, the method developed by S.M. Johnson is applicable (Table 7.4).

Table 7.4

	Time required		Lag	Time
Job	A	B	a	b
1	3	6	3	
2	7	2		2
3	4	7	4	
4	5	3		3
5	7	4		4

The method requires the lag time or the smaller time of the two machines to be calculated and entered under 'a' or 'b' according to machine A or B respectively (whichever has the shorter time). The sequence is determined by taking the lag time of the 'a' column first and arrange them in ascending order (in the example job 1, then 3). Then take the 'b' column and arrange them in descending order (in this case, jobs 5, 4 and 3). The sequence is therefore jobs 1, 3, 5, 4 and 2. The total elapse time can be calculated when the jobs are plotted on a time scale as in Fig. 7.3.

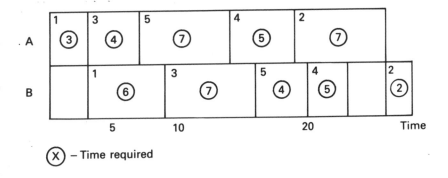

(X) – Time required

Fig. 7.3 Computation of total elapse time.

In the case of three machines, the two columns of lag time are calculated by taking the smaller of the total times of the first and second machines and the total times of the second and third machines. However, the condition that the minimum time on the first and third machines must be greater or equal to the second machine must be satisfied. The sequencing follows the same rule as the two-machine case.

In the example (see Table 7.5) the sequence is jobs 4, 6, 3, 1, 2 and 5, or jobs 4, 3, 6, 1, 2 and 5.

Table 7.5

Job	Time required A	B	C	A + B	B + C	Lag – Time a	c
1	10	4	9	14	13		13
2	8	3	8	11	11		11*
3	7	4	13	11	17	11	
4	5	2	7	7	9	7	
5	9	3	6	12	9		9
6	6	5	10	11	15	11	

* In case of tie, put it in the second column.

7.3.4 Minimization of total lateness

The following is an example of five jobs with their duration, deadline and penalty (if late) (Table 7.6). What is the best sequence from the table, so that total lateness will be minimized?

Table 7.6

Job	Production time (working days)	Due date (working days from start)	Penalty/day (£)
A	10	9	6
B	15	10	8
C	16	37	12
D	8	30	4
E	17	25	15

The solution procedure to this objective is as follows.

Step 1: Arrange jobs in increasing order of deadline (as in Table 7.7).

Table 7.7

Job	Due date	Prod. time	Cumulative Prod. time	Lateness	Penalty/Day	Total/penalty
A	9	10	10	1	6	6
B	10	15	25	15	8	120
E	25	17	42	17	15	25
D	30	8	50	20	4	80
C	37	16	66	29	12	348
Total		66		82		809

If all the jobs are finished before or on the due date, then the above sequence will be the optimal solution. Otherwise, proceed to Step 2.

Step 2: Arrange jobs in decreasing order of the reciprocal of the production time (as in Table 7.8).

Table 7.8

Job	Reciprocal prod. time	Due date	Cumulative prod. time	Lateness	Penalty/Day	Total penalty
D	1/8	30	8	–	4	–
A	1/10	9	18	9	6	54
B	1/15	10	33	23	8	184
C	1/16	37	49	12	12	144
E	1/17	25	66	41	15	615
Total				85		997

Step 3: Rearrange jobs one pair at a time, starting with the early job which is furthest right. Move this early job to the right of the pair. Carry on with this

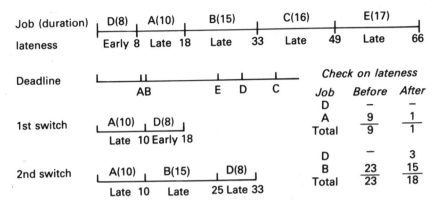

Fig. 7.4 Calculation of total lateness.

switching (whilst checking on the lateness for improvement) until all the jobs are late. To illustrate and work out this step, a time scale as in Fig. 7.4 is plotted. The optimum sequence is therefore A, B, D, C and E, which gives a total lateness of 72 days.

7.3.5 Minimization of total penalty

Using the same example in Section 7.2.4, the solution procedure to this objective is as follows:
Step 1: This is similar to Step 1 in the minimization of total lateness, and if all the jobs finish on or before the deadline, this will be the optimum sequence.
Step 2: Arrange jobs in decreasing order of penalty/production time.
   If all the jobs finish after the due date, this is the optimum sequence.

Table 7.9

Job	Prod. time	Cumulative prod. time	Prod. time	Due date	Lateness	Penalty/Day	Total penalty
E	0.882	17	17	25	–	15	
C	0.75	16	33	37	–	12	
A	0.60	10	43	9	34	6	204
B	0.53	15	58	10	48	8	384
D	0.5	8	66	30	36	4	144
Total		66			118		732

Step 3: Draw the time scale and rearrange one pair at a time and move the early jobs to the right, whilst checking the penalties to see if there is improvement. Continue until all the jobs are late.
   The optimum sequence indicated is A, E, C, B and D, and the total penalty is £636. (see Fig. 7.5).

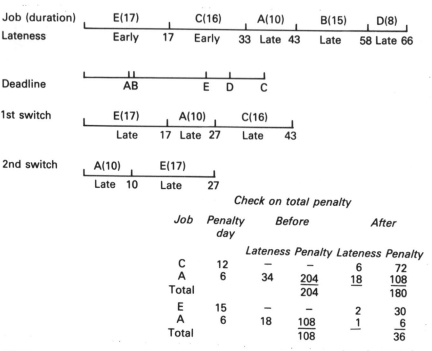

Fig. 7.5   Calculation of total penalty.

Job	Lateness	Penalty/day	Total penalty
A	1	6	6
E	2	15	30
C	6	12	72
B	48	8	384
D	36	4	144
			636

7.3.6 Conclusion

As has been shown, in the above examples, it is necessary to adopt some sort of priority rules in the allocation of resources, as well as in the sequencing of jobs. This is necessary in order to tackle the resource planning problem in a systematic manner and, at the same time, to give priority to activities on a rational basis, especially when there is competition for limited resource. The choice of priority will give different results for similar examples, therefore the decision as to which priority rules will be used must be one of judgement based upon experience. In property investment decisions, investors often have priority rules in allocating resources although the constraints for such decisions may be very different.

    The above examples have been used to illustrate some of the principles of resource planning and have involved the use of manual methods only. When problems are only slightly more complicated than those illustrated, the manual allocation of resources or the sequencing of jobs becomes out of the question. It is a tedious and complicated process which will lead to mistakes on the part of the allocator. It is in this field that the computer comes into its own; use of its facility to calculate quickly enables a quick review to be made not only of allocation in accordance with a specific priority rule, but also (if satisfactory results are not obtained) of taking a look at the allocation in the light of several different priority rules.

8 | Networks

8.1 CRITICAL PATH METHOD

8.1.1 Introduction

The real estate industry has been slow in taking up a wide range of quantitative techniques, but one area that is being developed and applied is the network planning method. The two network methods which are widely used are the **CPM** (Critical Path Method) and **PERT** (Programme Evaluation and Review Technique).

Although the bar or Gantt chart is more frequently used because of its simplicity in communicating short-term site programmes to operatives, it suffers from the defect of not showing the relationships between different operations from the management point of view. The critical path method (CPM) is better for the larger and more complex jobs, in that the network shows the interdependence of the various operations. Thus in the event of delay it is possible, from an examination of the network, to determine which operations are critical to prompt completion and to concentrate resources on these operations. A CPM network represents the sequence of operations or activities in a logical manner.

8.1.2 Fundamentals of CPM

The principal component of the CPM network diagram is the arrow, which represents an activity which is a time-consuming element in the programme. All projects can be broken down into a number of necessary activities. Each of these activities, which make up the whole project, is then represented by an arrow in the diagram.

An arrow not only represents the consumption of time, but also represents the consumption of certain resources such as labour, money and use of plant or materials. The length of the arrow on the diagram does not relate to the time which the activity takes or to the resources it consumes. The tail of the arrow represents the starting-point of the particular activity and the head represents the completion point. Examples of typical activities are shown in Fig. 8.1.

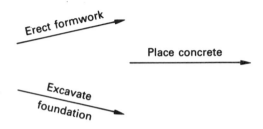

Fig. 8.1 Typical activities.

An event is the point at which an activity is completed or another activity commenced. It is an instantaneous point in time. These are usually indicated by a circle (or square) and can be an intersection of two or more activities as shown in Fig. 8.2.

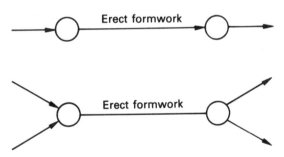

Fig. 8.2 Events and activities.

All networks are constructed logically on the principle of dependency; this is illustrated in Fig. 8.3.

Concrete cannot be poured until the formwork has been completely erected. Similarly, the curing of concrete cannot proceed until concrete has been poured. It becomes apparent that each activity must be preceded and succeeded by an event, and conversely, that each event must have an activity before it and an activity following it. The exceptions to this rule are at the commencement and the conclusion of a project. This will be illustrated later.

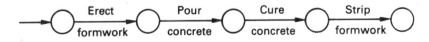

Fig. 8.3 Principle of dependency of activities.

8.1.3 Network logic

There are a number of abstract logical rules which are useful in the preparation of a network. If activities A, B and C are such that B and C must follow A, then it can be represented as follows:

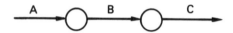

It can also be represented as:

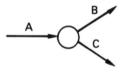

in which case it shows B and C are independent of each other.

If activity C follows B and activity D follows A, it could be represented as:

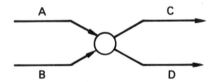

But the proper relation is:

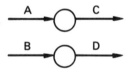

because no connection between C and A or B and D was stated.

Now, if both A and B must precede both C and D, the network is:

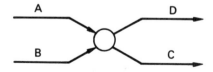

If A and B precede C, but only B precedes D, then it becomes difficult to represent this in the network. This can only be done by introducing a logical connection which is called a **dummy arrow** (or simply 'dummy') and is denoted by the broken arrow as follows:

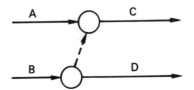

The network now shows that C follows A and B, but that D follows only B. The concept of the logical connection is indispensable to CPM.

Now consider a network example with two parallel chains of activities:

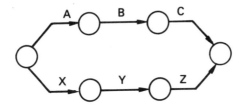

Now an activity P, originating at the first event, must precede both C and Y; the result is:

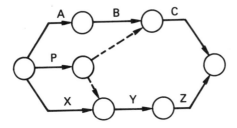

In the following network:

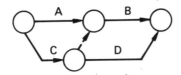

Adding terminal activity B which follows A, but is independent of C, is not accomplished by:

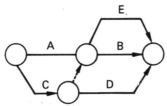

because it shows that E should follow C. In order to keep E independent of C, it can be shown as:

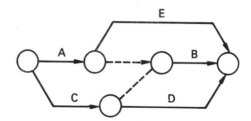

The above figure shows the introduction of another logical restraint after A. This is sometimes known as a **logic spreader**.

8.1.4 Network construction

The network is used to portray in graphical form the logic of the construction plan. Initially, it is used as a sequence plan as the time duration of various activities has yet to be answered.

In developing the network three questions must be asked of each operation:

(a) What activity must immediately precede this operation?
(b) What activity can immediately follow this operation?
(c) What activities can take place concurrently with this operation?

It is therefore useful to list all activities of the network and manipulate into a reasonably logical sequence. Answering the above questions for each activity will enable this manipulation to be carried out more easily.

Initial attempts at drawing the network diagram may be somewhat rough and may involve curved arrows going from left to right. Several arrows may cross one another. By arranging the diagram carefully, it is possible to tidy up the diagram and avoid too many situations of that kind.

8.1.5 An example

The nature of maintenance operation, namely the smallscale of jobs, diversity of the work content, non-repetitive nature of much of the work and lack of continuity, makes it difficult to permit the development of optimum methods. The use of network analysis is thus limited in the day-to-day routine maintenance operations. Networks tend to display their usefulness in maintenance project planning and in larger-sized maintenance jobs (e.g. in refurbishment projects). The following example is typical of the former category.

The Property Division of a certain corporation intends to convert some old terrace houses into a recreational centre on the ground level and residential apartments on the upper level. The activities which will take place during the pre-construction stage, together with their durations, are as follows:

	Activity	Duration (days)
A	Inception brief	3
B	Feasibility studies	6
C	Outline proposal	7
D	Planning approval	42
E	Scheme design	7
F	Sketch plan	12
G	Obtain finances	14
H	Building plan approval	30
I	Detailed design	10
J	Working drawings	14

To construct a network for the above activities it is helpful to first determine the activity dependency of all the activities. In other words, what are the activity (or activities) immediately preceding, following and running concurrently with each of the activities?

The **activity dependence table** for the example could follow this order:

Activity	Immediately preceding	Immediately following	Concurrent
A	–	B	–
B	A	C, D, F, G	–
C	B	E	D, F, G
D	B	H, I, J	C, F, G
E	C	–	D, F, G
F	B	–	C, D, E, G
G	B	–	C, D, E, F, H, I, J
H	D, E, F	–	G, I, J
I	D, E, F	–	G, H, J
J	D, E, F	–	G, H, I

Once the activity dependence table has been set, the network can be constructed (see Fig. 8.4). The network in the figure will help to illustrate to Maintenance Managers the sequence of work and the interdependency of the activities.

8.1.6 Time and network

After having constructed the network and ensured that the planning logic is sound, the scheduling process should be carried out by estimating the duration of each activity in the network (in the example this has been done). The estimated length of time required to carry out an activity is referred to as its **duration**.

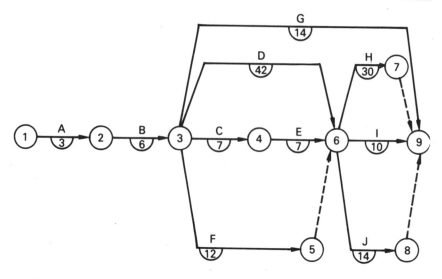

Fig. 8.4 Sequence of work and interdependency of activities.

Any convenient unit of time may be chosen in the scheduling process. This could be the hour, working day, calendar day, week or month. The most suitable units for construction work are the working day or working week. It is, however, important to distinguish between the working day and calendar day as deliveries often may be quoted in terms of calendar days and conversion from calendar days to working days must be made as necessary.

In estimating the duration of each activity it must also be assumed that normal and reasonable circumstances will apply to the work. It is important that the person responsible for carrying out the work has had a hand in this estimation. If all durations are estimated on the basis of normal activity, certain adjustments can be made at a later date when the whole network is reviewed because of any delay in the progress due to bad weather, accidents, etc.

The estimated activity duration is shown below the arrow of the relevant activity on the network diagram. Figure 8.5 shows the network diagram with the duration of the activities inserted in a semi-circle under the arrow.

8.1.7 Earliest event time

The earliest event time (Te) is the earliest time by which the event under consideration can be achieved. This can be established by considering all the previous activities merging with that event. Computation of the earliest event time is called the **forward pass**; it commences at the first event and ends in the last event.

Figure 8.5 illustrates the network for the example with the earliest event times superimposed upon it in the square box below each event.

Work is started at time zero and event 2 can be achieved at the end of the third day. The computation of earliest event times for events 3, 4 and 5 presents no problems. However, when considering the earliest event time for event 6 it is seen that this event may be approached through 3 paths 1–2–3–4–6; 1–2–3–5–6 and 1–2–3–6. Since 5–6 is a dummy, it bears no duration. Through the first path the earliest event time of 23 is arrived for event 6. Following the second path, the earliest event time for event 6 is 21. And following the third path, the *Te* for event 6 is 51. Therefore the earliest event time for event 6 is 51 since it is the earliest time at which all activities leading to it can be completed.

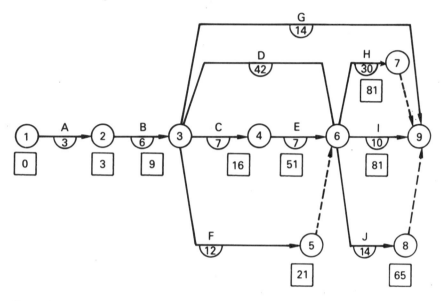

Fig. 8.5 Network diagram for the example.

The remainder of the diagram can be analysed in a similar way and it is seen from Fig. 8.5 that the project cannot be completed in less than 81 days.

8.1.8 Latest event time

The latest event time (T_l) is defined as the time by which a particular event must be achieved if there is to be no delay to the completion of the project in the overall duration. This phase of the computation is known as the **backward pass**.

The latest event time for the last event is taken as the earliest event time established by the forward pass computation. Thus event 9 is assigned the figure 81 and is placed in the triangular box adjacent to the earliest event time, as shown in Fig. 8.6.

The procedure for the computation of latest event times is simply a reversal of that for calculating the earliest event times. Events 8 and 7 thus have 81 as their latest event times since these are dummy activities. When considering

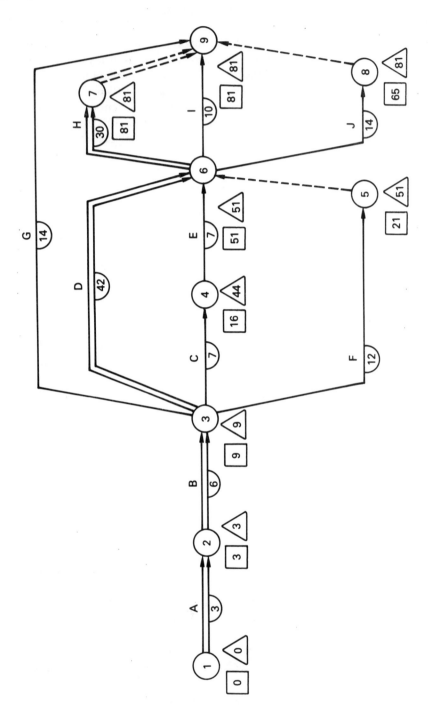

Fig. 8.6 The complete network diagram.

event 6, there are three merging paths through which the backward pass must be made. The first through 9–7, the second through 9–8 and the third direct from 9. Through the first path $T_l = 81 - 30 = 51$, through the second path $T_l = 81 - 14 = 67$ and through the third path $T_l = 81 - 10 = 71$. From the definition of latest event time it is clear that if event 6 is not completed by the end of 51st day the project cannot be completed in 81 days. Thus the latest event time for event 6 is 51.

The remainder of the diagram can be analysed in a similar manner and the latest event times for all events are shown in Fig. 8.6.

8.1.9 The critical path

On completing the earliest event times and the latest event times for the network, it is noticed that certain events have the same earliest event and latest event times. This means that any delay in the activities which connect such events will cause delay in the completion of the project. It can also be noticed that the path through these activities is the longest path through the diagram, and this is called the **critical path**. Each activity on this path is critical to the achievement of the project target.

Every network diagram will have a critical path; however, in certain cases there may be more than one critical path. From Fig. 8.6 it can be seen that the critical path is through events 1–2–3–6–7–9. It is useful to indicate this path on the diagram by thicker lines or by using a different colour, such as red, for the lines.

The criteria for finding the critical pass are thus,

(a) $Te = T_l$ for an event on the critical path;
(b) $T_lj - Tei - y = 0$ for a critical activity in which y is the duration of the activity.

Although we have mentioned earlier that activities connected by events which have their earliest time coinciding with their latest time will be on the critical path, this rule must still be tested by (b), above. In the example, activity 6–9 is not on the critical path.

In Fig. 8.6 if event 5 is examined, its earliest event time is 21, whilst its latest event time is 51. Thus a delay of 30 days in the activity 3–5 will not cause any delay in the overall completion time of the project. This leeway which an activity has is called a **float**; thus all activities which are not on the critical path will have floats.

For convenience in tabulation of activities with all the necessary information detailed, a format as shown in Table 8.1 is often used.

In this table the earliest event time associated with the event at the tail of any activity arrow is the earliest time by which the activity can commence. This is referred to as the **early start** (ES). If each activity proceeds from its **early start** time and takes the estimated duration against the arrow, it will have an **early finish** (EF). Thus, EF = ES + duration.

Similarly, if the event at the head of the arrow is examined, it will be seen that the **late finish** (LF) is the same as the latest event time for that event. The **late start** (LS) may be derived from the late finish by subtracting the duration from it:

$$LS = LF - duration$$

The float for any activity can be computed as:

$$Float = T_lj - Tei - duration$$

or, float $= LF - ES - duration$

Thus Table 8.1 shows the entire computation for the example.

Table 8.1

Activity	Duration (days)	Early Start	Early Finish	Late Start	Late Finish	Float
1–2	3	0	3	0	3	0
2–3	6	3	9	3	9	0
3–4	7	9	16	37	44	28
3–5	12	9	21	39	51	30
3–6	42	9	51	9	51	0
3–9	14	9	23	67	81	58
4–6	7	16	23	44	51	28
5–6	0	21	21	51	51	30
6–7	30	51	81	51	81	0
6–8	14	51	65	67	81	16
6–9	10	51	61	71	81	20
7–9	0	81	81	81	81	0
8–9	0	65	65	81	81	16

8.2 PERT

8.2.1 What is PERT?

PERT has as its central feature the critical path, but differs from the CPM in respect of job time estimates. PERT is concerned with situations in which job times are random variables. This may be due to the fact that the job time is simply unknown in the case of an entirely new project, or that the job times are influenced by uncontrollable variables – e.g., the weather or unknown construction details as in the refurbishment project.

8.2.2 Probability distribution of job times

Ideally, we would like to know the probability distribution of job times. Roughly speaking, this is a list of possible job times and the associated chances of occurrence of each time. The underlying probability distribution of job times is complicated in most instances and we require to know the type of probability distribution (e.g. discrete (as above), normal, gamma or beta) and the mean and variance of the distribution. It is rather unlikely that these parameters will be known in advance and PERT gives a simple, practicable means of obtaining reasonable estimates of mean and variance.

The person in charge of each job is asked to provide three estimates of job time: a 'pessimistic' estimate (a); an 'optimistic' estimate (b); and a 'most likely' estimate (m).

The arithmetic mean time (expected time) (e) can be approximated as a function of a, b and m, given the underlying distribution. For example, if the probability distribution is a beta distribution, the e is given by $er\ 1/6(a + b) + 2/3\ m$. The coefficients 1/6 and 2/3 are weights derived from the particular beta distribution which is usually more appropriate than the normal distribution because it has maximum and minimum limits. Once the mean time has been determined, this is used as in the CPM diagrams and analysis proceeds as for the CPM.

8.2.3 Limitations of PERT

There is serious difficulty involved in extending PERT to the whole network; because of the joining together of branches and the presence of jobs common to several paths, mathematical analysis becomes impossible.

Moreover multiple time estimates in network analysis are not fashionable nowadays because the approach is over-sophisticated in respect of many applications, and because eventually a single figure is employed. Unless the complications of PERT produce a significantly better mean figure, they will not be worth the cost and time involved. The choice of method should be determined by the particular problem at hand.

CPM itself can be used instead of PERT when job times are subject to variation. The first is what might be called the 'levels of criticalness' approach; in this, critical jobs now become 'first-level critical', and near-critical jobs become 'second-level critical'. For instance, in a building operation, jobs with three-days TF may be first-level critical. If all goes well, the second-level critical jobs would not determine the minimum completion time for the project. But if some of these job times depend on extraneous factors (e.g. the performance of a subcontractor), they may be 'unreliable' and become critical. If likely cases can be identified in advance, it would be useful to have some contingency plans worked out for speeding up some of the subsequent newly critical jobs.

Secondly, the acquisition of information costs both time and money. An economical approach to information gathering and critical path determination is as follows. For each job a rough-and-ready time study is done. With these approximate times, critical and near-critical jobs are identified. A detailed time study is then done only for those jobs. Time and money is thus saved by having only the rough study done for the 'far-from-critical' jobs.

Finally, if one wished to err on the side of caution, a fixed percentage could be added on to the initial time estimate for each job, so that although single figure estimates are then used, there is a built-in 'safety margin' for delays. This approach, along with the previous two, is one of approximation. The degrees of approximation are not specified, but the methods should not be dismissed. It is modification such as this that makes theoretical models more useful in practice.

8.2.4 Summary

CPM can help managers to identify the 'critical' activities which, if delayed, will extend overall project time unless compensatory reductions can be achieved in the durations of subsequent activities. Initially, activities are planned to start at the earliest time, but to provide a more uniform use of resources throughout the project it may be advantageous to delay the start of the non-critical activities. In this way, there is 'resource smoothing' or the even spread of resources.

However, the main advantage of CPM is that it demands the logical analysis of the proposed work and provides a means of identifying particular operations on which resources should be concentrated if for any reason the progress of the work is delayed. Both cost control and cost optimization can be pursued on the basis of CPM.

Although CPM has been traditionally applied to site works, it could equally well have been applied to the planning and arrangement processes and could provide an effective means of control where deadlines have to be met.

EXERCISES

Exercise 1

A small extension to a house would entail the following activities: Foundations, Partitions, Drainage, Pavings, Brickwork to Damp Proof Course, Roof, Fixing Windows, Brickwork to eaves, Ground floor, Opening through Existing wall, Plumber 1st and 2nd fixings, Electrician 1st and 2nd fixings, Glazing, Plastering, Joinery Fittings, Painting and Clean up. Draw a network to show the logical sequence and dependency of the activities.

Exercise 2

The refurbishment process of the building already mentioned could consist of the following activities and durations. Draw the network, determine the project duration and identify the critical path:

Operation/activity		*Duration (days)*
A	Preparation to commence work	5
B	Install floor strutting	4
C	Seal off chimney flues and fireplaces in basement	7
D	Dismantle timber staircase	7
E	Install raking shores to external walls to stabilize structures	5
F	Install needles and dead shores to walls to be demolished	6
G	Allow period for wall to take bearing on dead shore	3
H	Form new door opening in basement walls	2
I	Erect new brick piers on ground floor removing minimum amount of brickwork	3
J	Demolish existing brick wall and remove debris from site	4
K	Enlarge existing door opening and form new door openings to walls on ground floor	2
L	Negotiation with and appointment of Service Contractor	14
M	Electrician first fix	4
N	Erect new partitions on ground floor	2
O	Fix new beams (front)	2
P	Erect new door frame	2
Q	Fix new beams (back)	2
R	Build brickwork above new beam to underside of old work	2
S	Plumber first fix	5
T	Hang new doors and fix ironmongery	2
U	Finishings to new partitions	2
V	Allow time for brickwork to take bearing on new beams	7
W	Remove all shoring and make good needle perforations in walls	5
X	Make good finishings to match existing in new door openings formed, and flues and fire places sealed up	5
Y	Clean up	4
Z	Plumber second fix	2
AA	Electrician second fix	2
BB	Painting	3
CC	Handover	1

Resource management

9.1 Stock/inventory control
9.2 Deterministic stock problems
9.3 Queuing

9.1 STOCK/INVENTORY CONTROL

9.1.1 Introduction

An efficient stock/inventory policy is always an important requirement for the successful management of manufacturing and distributing enterprises. Any temporarily idle resource may be thought of as an inventory which has often been described as 'money in disguise'. Indeed, the stock may be of money itself, as in the case of holdings of cash, and it is possible to apply inventory models to cash management.

For the property manager, whatever material resources that are under his/her control can be considered as stocks; some of the most common examples are light fittings, lavatory supplies, cleaning aids, and so on. An efficient policy on the management of these items can help to save overall cost in the maintenance and upkeep of a building.

9.1.2 Reasons for holding stock

Broadly, there are four main groups of reasons for holding stock.

(1) As lot size inventories

These are economic reasons for holding stock and may be applied to all situations where inventories must be kept:

- To obtain quantity discounts. Larger lots may be ordered than are needed for immediate use in order that the benefits of large discounts are obtained.

- To reduce transport costs. The size of the lot ordered may be larger than immediately required, so that a full lorry load will be made and transport costs per item reduced.
- To keep ordering costs down. Instead of making two small orders, they may be combined into one and clerical and other administrative costs reduced.

(2) As buffer inventories

Here the reasons for holding stock are more applicable to the manufacturing industries, where stock control is an important aspect:

- To provide a buffer against fluctuation in the supply of an item and thus prevent a hold up of production. The buffer inventories are the raw materials that are used in the production.
- To provide a buffer against fluctuation in the supply of an item, so that the inability to supply normal customer demand does not arise. These refer to the finished goods of a manufacturer.
- To provide a buffer against fluctuations in the demand from customers (finished goods).
- To provide a buffer when production processes are carried out at different rates (semi-finished goods).

(3) As transit inventories

Transit inventories may be necessary as a temporary measure until men or machinery are available to move the items. The items while they are actually being moved may be called transit stock.

(4) As seasonal inventories

These are applicable to industries which produce goods that have seasonal demands:

- To keep production at a constant level and still allow for seasonal fluctuations in customer demand. If there are large seasonal fluctuations in demand and production attempts to follow them, the plant, labour and other resources used in production must be sufficient to cope with peak season demands and this is likely to be an uneconomic use of capital and other resources.
- To obtain the benefit of seasonal price fluctuations. Large quantities of materials may be purchased when the price is low.

9.1.3 Factors to be considered in stock control

Stock holding costs or the costs of carrying stocks

These costs are comprised of the following:

- Rent of premises for storage or the remuneration which would have been obtained if owned premises had been let or used for other purposes.
- Rates of storage premises.
- Heat and lighting of storage premises.
- Wages of warehouse staff.
- Cost of capital. The higher the stock, the more capital that is tied up. The rate of interest used may be taken as the cost of borrowing the money. A better rate (if higher) to use would be the highest return the firm could obtain if the capital tied up were invested elsewhere (or the opportunity cost of capital). The capital cost incurred will depend upon the total value of stock at various points in time and on the length of time the items are held in stock.
- Insurance. Most insurance of stock is based on an average level of stock held and is not likely to fluctuate very much unless there are large changes in the firm's average stock level.
- Depreciation. Some items deteriorate if kept longer than a certain period with the resultant wastage costs.
- Obsolescence. In some industries, fashions change rapidly and items in stock may have to be disposed of at a lower figure than the purchase price or production cost. This loss is a cost of holding stock.
- Spoilage and breakage.
- Movement costs. During storage, items may have to be moved around in the storage premises. This may be covered by additional staff, adding to the cost of holding stock.

Costs of replenishing stock

Replenishing stock requires the expenditure of money for the ordering for procurement of the stock. These costs vary with the order size and largely consist of the following:

- Cost of stationery and postage in sending for order.
- Clerical costs involved in sending order.
- Cost of transporting goods from supplier including postage for small items.
- Unloading costs.
- Costs of the item itself may vary with the order quantity. Discounts may vary with order quantity.

Costs of being out of stock (stockout) or penalty costs

These apply mainly to the manufacturing industry and may consist of one or a combination of the following:

- Cost of machines being idle. A production line may be stopped because the stock of raw materials has run out.

- Cost of machines being idle due to lack of a stock of spare parts.
- Backorder costs. If an item is out of stock, it may be necessary to write letters to inform the customer and this will involve postage and other administrative costs. This cost is likely to vary with the number of orders which cannot be met but may also be affected by the time orders are out of stock as information may have to be sent to customers to indicate that they have not been forgotten.
- If a customer's order is late, there may be penalties resulting from late delivery. This cost is likely to vary with the length of stockout.
- If a customer finds that the item he requires is out of stock, he may go to another supplier and the profit on this item will be lost. This type of cost or loss to the firm is difficult to quantify.
- If a customer finds that the item he requires is out of stock, he may be annoyed and the firm will lose some goodwill. He may not buy goods from the firm in the future, so that profit on all the items the customer would have bought from the firm will be lost. This intangible loss is impossible to calculate exactly.

Costs of operating an information system in connection with stock

These are:

- Actual stock count at periodic intervals which require time and labour costs.
- The cost of keeping records, either manually or by computer, and frequently updating these, so that information is reasonably up to date.
- The cost of assessing future demand, so that stocks are at the desirable level to meet demand.

9.2 DETERMINISTIC STOCK PROBLEMS

Of the various reasons given for stockholding, the most appropriate to the manager is that of economic lot size. As we have already mentioned, the property manager will be interested to know if the stock level of light bulbs, for example, is sufficient to meet the demands of the tenants at any one time. He or she will also want to know the economic order of quantity which balances the costs of holding stock and the costs of replenishing stock. It is with this objective that the basic but somewhat simplistic model is derived.

9.2.1 The Wilson formula

In most inventory systems, stock levels are reduced with time; they are restored by the arrival of a replenishment order and then the process is repeated. This is illustrated in Fig. 9.1.

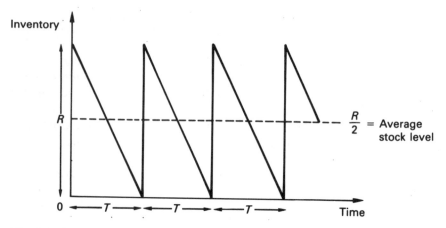

Fig. 9.1 Stock cycle.

The stock level in the figure diminishes at a uniform rate until it reaches zero. At this point, an order of size R is received, restoring the stock level to its maximum value. The length of time required for stocks to go from peak to peak (or equivalently from trough to trough) is one inventory cycle. We have thus assumed a known and predictable demand, together with immediate replacement (i.e. zero replenishment lead time). What is desired is the determination of the best value of the reorder quantity where best can be interpreted to mean 'that re-order quantity which allows operation at a minimum cost'. We do not aim to minimize costs per cycle which could be achieved by setting R = 0 or holding no stocks at all. (If this is the case, imagine having to purchase every single light bulb when an existing one blows.) Although we have listed several costs that could be incurred in inventory control, in this first model we shall consider only two types of cost, namely the cost of holding stock and the cost of replenishing stock.

Before we proceed to derive the model, let us summarize the assumptions:

1. There is only one product in stock.
2. All the stock is held in one location.
3. The demand for the product is constant over the period with which the problem is concerned.
4. The stock can be replenished immediately an order is placed.
5. The stock of the product must not be allowed to become negative.

Derivation of the formula

Let T = time period under review (T units of time);
Q = total demand in time period T;
R = size of replenishment order;
C = total cost of stockholding over time period T;

Cr = cost of placing an order (fixed cost of replenishment);
Ch = cost of holding a unit of stock for one unit of time.

Rate of demand per unit of time $= \dfrac{Q}{T}$

Average stock held $= \dfrac{R}{2}$

Number of orders placed in time $T = \dfrac{Q}{R}$

Total costs of holding stock over time $T = \dfrac{R \times Ch \times T}{2}$

(i.e. average stock × cost of holding one unit for unit time × time)

Total cost of placing orders for time period T:

$$\dfrac{Q \times Cr}{R}$$

i.e. number of orders placed × cost of placing an order)

The total cost of stock holding over time period T is:

$$C = \dfrac{R.Ch.T}{2} + \dfrac{Q.Cr}{R}$$

The components of total cost can be graphically presented (Fig. 9.2). From the graph in the figure, it can be seen that the total cost will be at a minimum where the stockholding costs equals the replenishing cost.

Equating the total cost of holding stock over time T to the total cost of placing orders for time period T, we have:

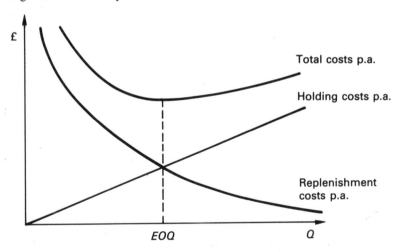

Fig. 9.2 Determination of EOQ.

$$\frac{RChT}{2} = \frac{QCr}{R}$$

$$R^2 = \frac{2QCr}{ChT}$$

$$R = \sqrt{\left(\frac{2QCr}{ChT}\right)}$$

Mathematically, to obtain the order quantity which will give the minimum total cost of stockholding we differentiate the following equation in respect to R and equate to zero and solve the new equation with respect to R:

$$C = \frac{R.Ch.T}{2} + \frac{Q.Cr}{R}$$

$$\frac{dC}{dR} = \frac{-QCr}{R^2} + \frac{ChT}{2} = 0$$

$$\frac{QCr}{R^2} = \frac{ChT}{2}$$

$$R^2ChT = 2QCr$$

$$R^2 = \frac{2QCr}{ChT}$$

$$R = \sqrt{\left(\frac{2QCr}{ChT}\right)}$$

It can be seen that algebraically both approaches produce the same formula. There are two other approaches to determine the *EOQ* which do not require the use of formulas. The first is the simplistic but logical trial-and-error approach. By computing at different levels of *EOQ*, we can find out the level which has the least total stockholding cost. This will then be the optimum *EOQ*. The second approach is by plotting the graphs of the cost and determine the point of intersection from the graph in Fig. 9.2.

EXAMPLES

Example 1

A maintenance department of a building estimated that the annual replacement of light bulbs has a mean of 500. Each bulb costs £5 and the holding cost is

15% of the unit cost. The ordering cost of each batch of bulbs is £200. Calculate the economic order of quantity assuming a uniform demand throughout the year:

$Q = 500$ per year

$T = 1$ year

$Cr = £200$

$Ch = £5 \times 0.15 = 0.75$ per item per year

Using the formula:

$$EOQ \text{ or } R = \sqrt{\left(\frac{2 \times Cr}{Ch}\right) \times \left(\frac{Q}{T}\right)}$$

$$= \sqrt{\left(\frac{2 \times 200}{0.75}\right) \times \left(\frac{500}{1}\right)}$$

$$= \sqrt{(266666.7)}$$

$$= 516.4$$

$$\text{Say, } \underline{516}$$

The total cost of holding stock is therefore:

$$C = \frac{R \times Ch \times T}{2} + \frac{Q \times Cr}{R}$$

$$= \frac{516 \times 0.75 \times 1}{2} + \frac{500 \times 200}{516}$$

$$= \underline{£387.3} \text{ per annum}$$

Quantity discounts

The derivation of the *EOQ* assumes a constant cost price per item, the only variable costs being the stockholding and delivery costs. However, it may be that a reduction in the cost price can be obtained if the items are bought in large enough quantities.

Example 2

A company requires 2500 units of an item per year for usage at a constant rate. The basic unit cost is £1 but discounts are given at 5% for orders of 1000–1999 and 7% for orders of 2000 or more. The fixed ordering cost is £10 per order. The annual inventory cost is 20% of unit price per year:

$Q = 2500$

$T = 1$ year

$Cp = £1$ (unit cost of purchase)

$Cr = £10$

$Ch1 = £1 \times 0.2$ (for orders less than 1000)

$Ch2 = £1 \times 0.95 \times 0.2 = 0.19$ (for orders of 1000–1999)

$Ch3 = £1 \times 0.93 \times 0.2 = 0.186$ (for orders greater than or equal to 2000)

For each different amount of orders, there will be a different replenishment quantity of EOQ since the holding cost is different.

The EOQs for the three different discount ranges can be calculated as follows:

$$R1 \text{ (or EOQ1)} = \sqrt{\left(\frac{2Cr}{Ch1}\right) \times \left(\frac{Q}{T}\right)}$$

$$= \sqrt{\left(\frac{2 \times 10 \times 2500}{0.2}\right)}$$

$$= \underline{500}$$

$$R2 \text{ (or EOQ2)} = \sqrt{(2 \times 10 \times 2500)/0.19}$$

$$= \underline{513}$$

$$R3 \text{ (or EOQ3)} = \sqrt{(2 \times 10 \times 2500)/0.186}$$

$$= \underline{519}$$

To compare each of the different discount offers, the total cost of holding stock needs to be calculated.

For No Discount:

$$\text{Total cost} = \frac{R1 \times Ch1 \times T}{2} + \frac{Q \times Cr}{R1} + (Cp \times Q)$$

$$= \frac{500 \times 0.2 \times 1}{2} + \frac{2500 \times 10}{500} + (1 \times 2500)$$

$$= \underline{2600}$$

For 5% Discount,

$$\text{Total cost} = \frac{R2 \times Ch2 \times T}{2} + \frac{Q \times Cr}{R2} + (Cp \times Q)$$

To benefit from the discount, the minimum replenishment order $R2$, must be at least 1000 (instead of 513 as calculated):

$$\frac{1000 \times 0.19 \times 1}{2} + \frac{2500 \times 10}{1000} + (1 \times 2500 \times 0.95)$$

$$= 95 + 25 + 2375$$

$$= \underline{2495}$$

For 7% Discount:

$$\text{Total cost} = \frac{R3 \times Ch3 \times T}{2} + \frac{Q \times Cr}{R3} + (Cp \times Q)$$

Similarly, to benefit from the discount the minimum order, $R3$, must be at least 2000 (instead of 519 as calculated):

$$\frac{2000 \times 0.186 \times 1}{2} + \frac{2500 \times 10}{2000} + (1 \times 2500 \times 0.93)$$

$$= 186 + 12.5 + 2325$$

$$= \underline{2523.5}$$

Therefore the optimum EOQ is 1000 units.

Lead time

So far, the model for establishing the EOQ has assumed zero lead time (i.e. instantaneous replenishment). However, in practice, there is usually a delay between the time of placement of an order for replenishment of stock and the time of arrival of the goods in inventory; this is called the **lead time**. The effects of this change of assumption are minimal – the order for replenishment must simply be placed when the amount of inventory falls to the level of lead time demand, which is the re-order level R. This is shown in Fig. 9.3.

Example 3

A builder's merchant sells 26 000 m of pipe lagging per year. The cost of lagging is £3 per metre. Holding cost is 12% of the average stock. Ordering

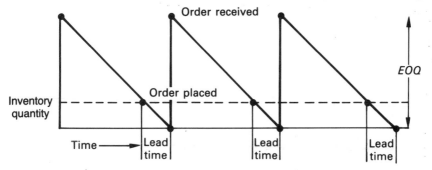

Fig. 9.3 Inventory with constant usage and constant lead time.

cost is £100 per order. Assuming a lead time of 1 week and a constant demand, calculate:

(a) the economic order of quantity;
(b) the re-order level.

The assumptions of the earlier examples are applicable to this case as well:

(a) $\text{EOQ} = \sqrt{\left(\dfrac{2 \times 100 \times 26\,000}{0.12 \times 3}\right)}$ where $Cr = 100$

$$= 3801 \qquad\qquad\qquad Ch = 0.12 \times 3$$

$$Q = 26\,000$$

$$T = 1$$

(b) The re-order level $= 26\,000/52 = 500$ since lead time is 1 week.

Safety stock

So far in the examples, we have assumed constant demand and lead time. These problems can be dealt with by the inventory models which provide us with a closer approximation to reality.

However, the most obvious problem of the inventory models is that certainty does not exist in most inventory situations. Both acquisition lead time and the demand for items usually fluctuate in a manner not completely known in advance to the decision- maker. Demand for the inventory items can be greater or less than anticipated due to external and internal factors such as weather change and power failures. Similarly, the acquisition lead time can also vary from favourable to unfavourable due to the supplier(s) and/or the transport-ation carrier(s).

If inventory is not available when needed due to any internal or external factor, a stockout occurs. This situation can lead to a decrease in profits and possibly losses. Figure 9.4 shows the problem of no stock when needed. It should be noted that the inventory level does not return to its original point as in Fig. 9.3 since back orders must be filled.

The re-order point is defined as a condition that signals a purchase order should be placed to replenish the inventory stock of some item. Thus the two variables (demand and lead time) are an integral part of the re-order point. The computation for the re-order point is the result of multiplying demand, expressed in terms of number of units per day, times the lead time in days. However, what must a firm do to provide for stockouts? The calcula-tion for the re-order period must be adjusted to provide for stockouts, result-ing in the addition of safety stock or buffer stock to the above computation; thus:

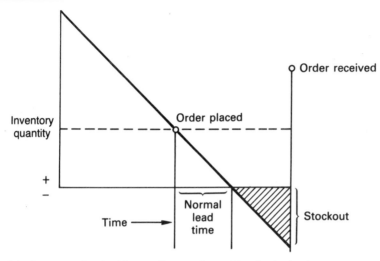

Fig. 9.4 Inventory level with no safety stock resulting in stockout.

Re-order point = average daily usage × lead time in days + safety stock

The term 'safety stock' refers to extra inventory held as a protection against the possibility of a stockout. In the basic model, stocks are never needed since by assumption nothing is unforeseen. However, if safety stocks were added in, they would represent a level below which inventory would never fall. Both the average level of stock and the re-order level are shifted up by the amount of safety stock. The situation is illustrated in Fig. 9.5.

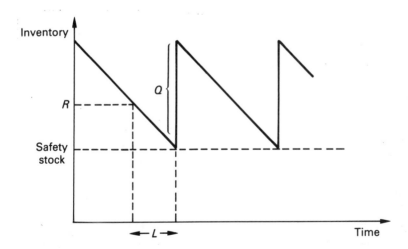

Fig. 9.5 Safety stock.

Example 4

The first step, utilizing the probability approach, is to analyse the past records for the inventory item in order that a probability percentage can be assigned to the various quantities of usage during the re-order period. The approach can be best illustrated by an example.

The ABC Manufacturing Co. has compiled data for an item of stock (Table 9.1). The company has found the economic ordering quantity to be 250 units with an average daily usage of 5 units. Lead time for this item is 21 days. Based upon the data in Table 9.1, the firm could re-order 250 units when the level of its stock falls to 105 units (5 (average daily usage) × 21 (lead time in days)), but it will be out of stock 8% of the time (0.06 + 0.02). What should management do about this figure of 8% stockouts? The answer may or may not be a certain amount of safety stock. A procedure for determining safety stock is needed. Obviously, management desires to pick that level of stock which will yield the lowest total cost for stockouts and inventory carrying costs of safety stock. Since the firm's re-order point is 105 units, the following safety stocks are considered: 5 units for a usage of 110 units, and 10 units for a usage of 115 units. Five units of safety stock would cover a usage of 110 units during the re-order period, resulting in the firm being out of stock 0.02 of the time. Ten units of safety stock would take care of all usage during the re-order period, thereby never having a stockout occur.

Table 9.1 Probabilities of usage during re-order period

Usage during re-ordered period (units)	Number of times this quantity was used	Usage probability
90	7	7/100 or 0.07
95	10	10/100 or 0.10
100	25	25/100 or 0.25
105	50	50/100 or 0.50
110	6	6/100 or 0.06
115	2	2/100 or 0.02
	100 times	1.00

The next step is to construct a table reflecting for each level of safety stock the total annual stockout costs. In order to do this, the cost of being out of stock for each unit must be calculated. In our example, the cost is £30 per item. Also consideration must be given to the number of times per year the company re-orders since a firm will be in danger of running out of stock that many times during the year. In the example, the EOQ formula indicates that five orders per year is optimum. The costs of being out of stock for each level is shown in Table 9.2.

The final step after determining the total annual stockout costs is to calculate the annual carrying costs per year. In this example, the cost per year of carrying each item in inventory is £4.00. Table 9.3 gives total costs of safety stock. The lowest total cost in this table is £35 for a safety stock of 5 items. The present re-order point of 105 units must be increased to provide for the safety stock of 5 units. Thus the re-order point is 110.

Table 9.2 Costs of being out of stock

Safety stock	Probability of being out of stock	Number short	Expected annual cost (No. short × probability of being short × cost of being out per unit × No. orders/yr)	Total annual stockout costs
0	0.06 when use is 110	5	$5 \times 0.06 \times £30 \times 5 = £45$	
	0.02 when use is 115	10	$10 \times 0.02 \times £30 \times 5 = £30$	£75
5	0.02 when use is 115	5	$5 \times 0.02 \times £30 \times 5 = £15$	£15
10	0	0	0	£ 0

Table 9.3 Costs of safety stock

Safety stock	Cost of being out of stock	Annual carrying cost (No. carried × cost)	Total cost/year (stockout cost plus carrying Costs)
0	£75	0	£75
5	£15	$5 \times £4 = £20$	£35
10	£ 0	$10 \times £4 = £40$	£40

Summary

The EOQ concept has proven widely useful in practical inventory control but the models presented here, for the most part, would require elaboration to address complex practical problems. For property investment decisions, the concept may be limited in application but if these were to be viewed in a larger context, for instance incorporating property management issues, there would be greater scope for application.

9.3 QUEUING

9.3.1 Introduction

Queues are a commonplace experience, for example:

- people waiting in shops and at service counters;
- cars waiting at traffic lights;

- aircraft waiting to land;
- ships waiting to enter port.

There are other examples of queues which may be less obvious but are nevertheless queues in principle; for example:

- goods waiting in inventory;
- telephone subscribers waiting for a clear line;
- papers in an in-tray;
- bank loans waiting to be cleared.

9.3.2 Queuing theory

Queuing theory, or as it is sometimes called the study of waiting line problems, is, as the name implies, a study of the statistical description of the behaviour of queues.

In some systems, it may be desirable to eliminate queues. This might be the case in the casualty department of a hospital. In other cases, though it may not be necessary to eliminate the queue completely, it may be desirable to keep the number in the queue below a certain limiting figure. An example of this form might be a queue at traffic lights where too long a queue would result in the obstruction of other roads.

In some systems, it may be desirable to have a queue for all or most of the time, even though it may be desirable to keep the actual length down below a certain figure. This type of queue might occur where items were being processed by a piece of machinery which was on hire at a very high hourly rate, so that the maximum use of the machine's time should be aimed for.

The items coming into the queue system are generally called customers, though they need not necessarily be people. They might include lorries waiting to be filled by an excavator or skeleton cores for doors coming off one type of machine and waiting to be processed on another machine. In property management problems, one or two men may be available to repair a very large number of service equipment and, in this case, the equipment which has broken down can be regarded as customers waiting to be serviced by the mechanics.

The items which cause the hold-ups are called the servers or service units or, in some cases, the service stations. In the above examples, the servers are the excavator, the machine carrying out the process on the door cores and the mechanics respectively.

9.3.3 Types of queuing model

Queuing problems may be investigated when some or all of the following occur:

- It is possible to control the flow of customers.
- It is possible to control the number of service units or the time taken for service.
- It is possible to control the order in which customers are served.
- Costs are associated with the idle time of service units.
- Costs are associated with the time spent in the queue by customers; these costs may, in some circumstances, be difficult to quantify and may in the case of a supermarket be an assessment of the loss of customers due to long waiting periods. In the case of lorries hired by a contractor, the time spent waiting would have to be paid at the hire rate.

In order to specify fully a queuing problem, it is necessary to state the following characteristics:

- The arrival pattern of customers.
- The service pattern.
- The queue discipline.

9.3.4 Arrival pattern

The arrival pattern of customers may be a regular one where the interval between the arrival of customers is constant throughout the period of operation of the system. If the arrival is random and the probability of two or more items arriving at any instant of time is zero, then the number of arrivals in unit time may have a Poisson distribution. The arrival time between customers may also have other forms of distribution including the normal, but only the first two arrival distributions will be considered here. Customers may also arrive in batches instead of singly.

9.3.5 Service pattern

The service pattern (Fig. 9.6) will be determined by:
(a) The number of service points a customer must pass through:

- If there is only one service point and all customers must pass through it, then this is called a **single channel**.
- If there are several points but all customers have to pass through all of them, as in the case of some production systems, then this is called a **line** though it is still a single channel or service line, single channel.
- If there are several points but each customer only has to pass through one then this is a **multi-channel** situation.
- If there are several lines of service but customers only have to pass through one of these lines then this is also a multi-channel situation.

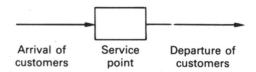

Fig. 9.6 Single point, single channel.

(b) The statistical distribution of the service time:

- The service time for each customer may be the same. If the arrival time and the service time are both constant, the queue is called **determinate** (or deterministic).
- The service time may be a random variable and the most common form of distribution assumed to be applicable, in this case, is the **exponential distribution**.
- Other types of distributions are included in some models, and these include the Erlang distribution and the normal distribution. In some cases, the length of service time may vary with the length of the queue.

(c) The number of customers served by a service point at the same time. In most cases, customers are served one at a time, but in others they may be served in batches. For example, a bus may not start (serve) until all the seats have been occupied.

(d) The time when service will be available. The service may stop at intervals; in the case of a building-site, this may be for a lunchbreak or teabreak or the end of the day.

9.3.6 Queue discipline

The queue discipline defines what happens between the time a customer arrives wanting service and the time when service actually starts. Customers may form a single queue for service and first to arrive is served first (first come, first served). If there is more than one server, there may still only be one queue, and as soon as a server becomes available, the first person in the queue may go to him.

If there are several servers, then separate queues may be formed for each server. In this case, a customer might go to the queue which is the shortest when he arrives. The choice of queue may depend upon the service point characteristics. The position of a service point may also influence the choice of queue which a customer enters. In a large supermarket a customer may go to a checkout point near the exit on to the main shopping street rather than one near a rear exit on a side street.

In some cases, customers may change from one queue to another (jockey). Where there are multiple channels and queues some of the customers will move from what appears to be a slow-moving queue to a faster-moving one, and if this new one should slow, then they may move back again. Some customers may see the length of a queue and leave (balking), or join the queue and then leave before being served (renege). In some circumstances, the queue length may be restricted and any customers arriving when it has reached the maximum length will be sent away.

The order in which customers are served also forms part of the queue discipline. One order has already been mentioned, first in, first out (FIFO): often this is regarded as a fair method of dealing with customers, when customers are actual people. Last in, first out (LIFO) is also sometimes used but this rarely occurs.

Priority of a customer dependent upon the time taken to carry out the service may be used, and the cost of a customer waiting may also be used as a criterion for choosing priorities. The urgency of a particular job may form a customer priority in the case of queuing problems relating to production.

Random selection of customers from the queue may occur, as in the case of a telephone operator who is unable to tell which waiting call was the first to arrive at the switchboard. The status of a particular customer may determine the priority.

There are a variety of complex situations which can occur in queuing problems, but there will not be space here to tackle all of them.

9.3.7 Simple queues

Assumptions and Parameters

There is a surprisingly large number of conditions that need to be satisfied for a so-called 'simple queue situation'. Precisely how many depends on how they are grouped together. They may include:

- Random arrivals and service.
- Large population of indivisible customers.
- Single queue, unlimited capacity.
- FIFO queue discipline, no reneging or balking.
- Single service point.
- One-at-a-time service.
- Average rate of service greater than average rate of arrivals.

Notation and formulae

There is a standard notation in queueing theory:

λ = mean rate of arrivals;
μ = mean rate of service.

When arrivals are purely random, the expected (in a statistical sense) number of arrivals in a 'short' time period is given by the Poisson distribution:

$$P(n \text{ arrivals in time interval} T) = \frac{(\lambda T)^n \exp(-\lambda T)}{n!}$$

For instance, if $\lambda = 0.4$ per second, the probability of three arrivals ($n = 3$) in a two-second interval ($T = 2$) would be:

$$P(3 \text{ in } 2) = (0.8) \, 3 \exp(-0.8)/6 = 0.03834$$

If arrivals follow a Poisson distribution, the distribution of time between arrivals is negative exponential. It is written as:

$$f(T) = \lambda \exp(-\lambda T)$$

where $f(T)$ is a density function.

The probability of up to T seconds elapsing between arrivals is the proportion of the total area under the curve $f(T)$ that lies between 0 and T. This calculation requires integration and will not be discussed here.

More important are the system parameters that can be obtained if a simple queue situation exists. The most commonly used of these measures are:

Traffic intensity:

$$\rho = \frac{\lambda}{\mu} \tag{9.1}$$

Probability of system containing n customers:

$$P_n = \rho^n (1 - \rho) \tag{9.2}$$

Probability that there are at least N customers in the system:

$$\rho^N \tag{9.3}$$

Average number of customers in the system:

$$\frac{\lambda}{\mu - \lambda} \tag{9.4}$$

Average length of queue (number of items in queue):

$$\frac{\lambda^2}{\mu(\mu - \lambda)} \tag{9.5}$$

Average length of queue (excluding zero queues):

$$\frac{\mu}{\mu - \lambda} \tag{9.6}$$

Average system process time (ASPT):

$$\frac{1}{\mu - \lambda} \tag{9.7}$$

Average queuing time:

$$\frac{\lambda}{\mu(\mu - \lambda)} \tag{9.8}$$

Average number of customers being served:

$$\frac{\lambda}{\mu} \tag{9.9}$$

The traffic intensity is a measure of use of the system and in a simple queue situation must be less than one. ρ is also the probability of an arrival having to wait. In other words, it is the probability that there are one or more customers already in the system. Thus, since the system must be in some state (must contain some number of customers), $1 - \rho$ is P_0, the probability that there are no customers in the system at any time. All the other states of the system (containing 1, 2, ..., n members) are related to ρ by equation (9.2). Equation (9.3) results from the summation from N to ∞ of the ρ in equation (9.2).

The most important system parameters are equations (9.4)–(9.9) and we shall illustrate the use of these by examples.

On average, cars arrive at a petrol station every 3 min. The single attendant is capable of serving on average 30 cars per hour. Service times and inter-arrival times follow a negative exponential distribution:

(i) What is the probability of a car arriving and having to wait for service?
(ii) What is the probability of a car arriving and finding at least one car already at the petrol station?
(iii) What is the average number of customers at the garage at any moment?
(iv) What is the length of time that a customer would expect to spend at the garage?
(v) What is the average number of customers in the garage who are not being served?
(vi) The garage proprietor considers that unless customers can expect to be served immediately on arrival 40% of the time, trade will eventually drop off. With the original demand pattern, what percentage reduction in service time would be necessary to achieve this result?

If, on average, a car arrives every 3 min, this means an average of 20 per hour; so choosing an hour as the basic time unit, $\lambda = 20$. Clearly, $\mu = 30$. As the service times and arrival times follow a negative exponential distribution, simple queue conditions prevail. The answers to the above questions can be easily calculated:

(i) This is $\rho = \dfrac{\lambda}{\mu} = \dfrac{20}{30} = \dfrac{2}{3}$

(ii) This is also $\rho = \dfrac{2}{3}$

(iii) This is $\dfrac{\lambda}{\mu - \lambda} = 2$

(iv) This is the average system process time:

$$\frac{1}{\mu - \lambda} = \frac{1}{10}\,h, \text{ or } 6\,\text{min}$$

(v) Assuming that there are no general hangers-on, this means that the average number in the queue:

$$\frac{\lambda^2}{\mu(\mu - \lambda)} = 4/3$$

(vi) The reduction in service time must be such as to produce $P_0 = 0.4$. A person will be served immediately only if there is no one already there. Thus $P_0 = 1 - \rho = 0.4$, so that $\rho = 0.6$. With a given arrival rate of $\lambda = 20$, we therefore require that $20/\mu = 0.6$, so that $\mu = 33\frac{1}{3}$.

This value of μ implies a service time of $1/33\frac{1}{3} \times 60\,\text{min} = 1.8\,\text{min}$, which is a 10% reduction on the original time.

9.3.8 Multiple service channels

Generally

There are many ways in which the rigid conditions of a simple queue can be relaxed. One of these concerns the number of service points that are operated. We shall now consider the consequences of having several service points instead of one. It will be assumed that these service points are approached by one queue only – the customer at the head of the queue going to the next service point or server who becomes free. This kind of arrangement, often seen in banks, gives the best theoretical performance and we shall endeavour to explain why multiple queues are nevertheless also commonly observed.

Assumptions and parameters

First, the single queue, multi-server case: we shall assume that there are no other changes from the simple queue situation and the distribution of service time at each point is (the same) negative exponential distribution. New formulae will be required here; it is more convenient to write these in terms of p. The main results, with c identical service points, are:
Probability of a customer having to wait for service:

$$= \frac{(\rho c)^c}{c!(1 - p)}\, P_0 \tag{9.10}$$

average number of customers in the system:

$$= \frac{\rho(\rho c)^c}{c!(1-\rho)^2} P_0 + \rho c \qquad (9.11)$$

average number of customers in the queue:

$$= \frac{\rho(\rho c)^c}{c!(1-\rho)^2} P_0 \qquad (9.12)$$

average time a customer is in the system:

$$= \frac{(\rho c)^c}{c!(1-\rho)^2 c\mu} P_0 + \frac{1}{\mu} \qquad (9.13)$$

average time a customer is in the queue:

$$= \frac{(\rho c)^c}{c!(1-\rho)^2 c\mu} P_0 \qquad (9.14)$$

in which:

$$P_0 = \frac{c!(1-\rho)}{(\rho c)^c + c!(1-\rho) \sum_{n=0}^{c-1} \frac{1}{n!} (\rho c)^n} \qquad (9.15)$$

and where, most importantly, it should be noted that:

$$\rho = \frac{\lambda}{c\mu} \qquad (9.16)$$

Note that ρ is defined as λ over $c\mu$. We see that our original ρ was a special case where $c = 1$. All of the formulae involve P_0 which is the probability that there are no customers in the system. Equation (9.10) is the probability that there are $c\mu$ or more customers in the system (i.e. all the service points are occupied). This is the only circumstance in which waiting is necessary. It is usually best to begin any workings by calculating P_0, and to minimize rounding error it is best to leave things in fractional form until the very end.

EXAMPLES

Example 1

A bank has arranged its services such that customers requiring cash only are served from any one of three service points, there being only one queue. The average rate of arrivals is 72 per hour. Each cashier takes on average 2 min to serve a customer. The bank is considering the installation of an automatic cash machine which would mean that there would be only one service point which would take on average only 40 s to serve each customer. Which system produces the lesser average system process time?

We observe that $\lambda = 72$ and $\mu = 30$ at each service point; thus:

$$\rho = \frac{\lambda}{c\mu} = \frac{72}{3(30)} = 0.8$$

Now find P_0

$$P_0 = \frac{3 \times 2 \times 1(1 - 0.8)}{(0.8 \times 3)^3 + 3 \times 2 \times 1(1 - 0.8)\left[\frac{1}{1}(2.4)^0 + \frac{1}{1}(2.4)^1 + \frac{1}{1}(2.4)^2\right]}$$

Thus:

$$P_0 = \frac{1.2}{21.36}$$

Therefore,

$$ASPT = \frac{13.824}{6(0.2)^2\,90} \times \frac{1.2}{21.36} + \frac{1}{30}\,h$$

$$= \frac{16.5888}{461.376} + \frac{1}{30}$$

$$= 0.069288\,h, \text{ or } 4 \text{ min } 9 \text{ s}$$

The probability of a customer having to wait for service is:

$$P(n \geqslant c) = \frac{13.824}{1.2} \times \frac{1.2}{21.36} = 0.6472$$

Now consider the proposed new arrangement. This is a simple queue situation with $\lambda = 72$ as before but with $\mu = 90$. We could use the new formulae with $c = 1$ but it is much simpler to remember that:

$$ASPT = \frac{1}{\mu - \lambda} = \frac{1}{90 - 72}\,h = 3\text{min } 20 \text{ s}$$

So the new arrangement is preferable. But under the new circumstances the probability of having to wait is 0.8, which is greater. Under the old arrangements, there was more chance of a customer getting served right away.

Perhaps the comparison is not so clear-cut as it first appeared, especially as we are not considering the 'personal factor' and the fact that queuing time may be psychologically more significant than overall process time. These factors should be borne in mind in the practical situation.

In this problem, we contrasted the performance of three service points and one queue, with one queue and one service point working three times as fast.

Another possibility would be to have three service points, each with its own separate queue. To evaluate this last arrangement start by assuming no jockeying, no reneging, and no balking with random arrivals at the rate of $\lambda/3$ into each queue. We specifically assume that the rate of arrivals into each queue is not state dependent, namely that a new arrival is equally likely to join any of the queues regardless of their relative lengths. This being the case, we have the arrangement:

$$\lambda = 24 \xrightarrow{\quad} \boxed{\mu = 30}$$

$$\lambda = 24 \xrightarrow{\quad} \boxed{\mu = 30}$$

$$\lambda = 24 \xrightarrow{\quad} \boxed{\mu = 30}$$

Clearly, for each of these three simple queue situations:

$$ASPT = \frac{1}{30 - 24} = \frac{1}{6} \, \text{h} = 10 \, \text{min}$$

which is by far the worst performance of the three arrangements. The following ordering (by ASPT) is therefore generally true:

- single queue single server operating at rate $n\mu$;
- single queue n servers operating at rate μ;
- n queues n servers operating at rate μ.

Example 2

The average rate of arrivals at a self-service store is 30 per hour. At present there is one cashier who, on average, can attend to 45 customers per hour. The store proprietor estimates that each extra minute of system process time per customer means the loss of 10p profit. An assistant can be provided for the cashier (to weigh and wrap goods, etc.) and, in these circumstances, the service point can on average deal with 60 people per hour. The wage rate of the assistant would be £3 per hour. Is it worth taking on the assistant?

$$\lambda = 30, \mu = 45$$

thus the average system process time originally is: $1/(u - \lambda) = 1/15$ hour per customer, so that there is an opportunity loss per customer of $1/15 \times 60 \times 0.10 = £0.40$; but there are 30 customers per hour to consider, so that the store's lost profit per hour is $£10.40 \times 30 = £12$ per hour with the original arrangement. With the assistant provided, $ASPT = 1/(60 - 30) = 1/30$ hour per customer, so that the opportunity loss per hour is $1/30 \times 60 \times 0.10 \times 30 = £6$, a reduction of £6 on the original figure. If we now deduct the assistant's wages, there is a net saving of £3 per hour, so that it is definitely worthwhile having the assistant.

The question arises as to where the 10p per minute cost comes from. Such figures are difficult to assess. In such cases, it is advisable to conduct a sensitivity analysis on the result. What is the least value of the lost profit per minute for which the assistant would be worthwhile? This is L, such that:

$$\frac{1 \times 60 \times L \times 30 + 3}{30} \leqslant \frac{1 \times 60 \times L \times 30}{15}$$

i.e. $60L + 3 \leqslant 120L$

$60L \geqslant £3$

$L \geqslant £0.05 \text{ or } 5 \text{ p}$

It may be that while the manager was by no means sure that 10p per minute was the correct figure, he may be well satisfied that it is in excess of the required minimum of 5p.

Example 3

A maintenance section receives on the average 8-hour working day 8 work orders (or 1 per hour according to a Poisson distribution). To answer each request, the section has a team of 4 men which will take 3/4 h per job (including travelling time within the building) and this service time is exponentially distributed. The management feels that in order to improve the service to the tenants, the average waiting time must be cut down. They are considering splitting the 4-man team into two 2-man teams which will take double the time to service each request. Is there an improvement with this new system?

In the original case, $\lambda = 1$ and $\mu = 1.33$. The average queuing time is therefore:

$$\frac{1}{\mu(\mu - \lambda)} = \frac{1}{1.33(1.33 - 1)} = 2.278$$

In the new arrangement, $\lambda = 1$ and $\mu = 0.667$. The average queuing time is therefore:

$$\frac{(\rho c)^c}{c!(1 - \rho)^2 \, c\mu} P_0 = 1.93 \text{ h}$$

where $\rho = \dfrac{1}{2 \times 0.667} = 0.7502$

and $P_0 = 0.14274$

Therefore, in terms of queuing time, the new arrangement is preferable.

Summary

There are a great number of different queuing models, each with its own formulae and results. Some of the main variations are:

1. Capacitated systems, with limits on the numbers allowed in the queue.
2. Different service time distribution, for example, the Erlang (gamma) distribution or indeed any arbitrary distribution.
3. Queues in networks.
4. Transient situations (what happens if a system never has time to settle down?).
5. Priority systems (different classes of customers, as in computer usage).

Queuing problems, in practice, can be extremely complex. Analytical solutions may be impossible, and if experimentation in the real situation is ruled out, then the only scientific alternative is simulation. In cases where the analytical methods can be used, they are extremely valuable. In any case, a study of the analytical methods helps to develop a 'feel' for the problem.

As with most quantitative techniques, queuing theory is hardly applied in property investment decision-making. There are, however, situations in property management where this technique may be of use as a management tool. In any case, the modern property manager should be equipped with such tools as clients become more demanding.

EXERCISES

Exercise 1

An office of a planning department has members of the public arriving with problems on an average of 10 per hour. The staff can serve only one person at a time, but they have a potential service rate of 20 per hour. Assuming the arrivals are of a Poisson distribution and service times are exponentially distributed, find:

(a) the expected number of the members of the public in the office either being served or waiting;
(b) the average number waiting for service;
(c) the average time a member of the public spends waiting in the queue;
(d) the average time a member of the public spends in the office.

Exercise 2

On average, 96 customers per 24-hour day require services in an emergency service firm. Also, on the average, the customers require 10 min of active attention:

(a) Assuming that the emergency service can only handle one emergency at a time, describe the degree of congestion.

(b) If it costs £100/customer service to obtain a servicing time of 10 min, and that each minute of decrease in this average time would costs £10 per customer service, how much per customer would have to be budgeted for by the firm to decrease the average size of the queue to half a customer?

(c) What would the servicing rate have to be if the chance of two or more customers either being serviced or waiting at the same time was to be kept below 10%?

The analysis of investment risk | 10

10.1 INTRODUCTION

All investment decisions are concerned about future events, all of which by their very nature are uncertain. When future events are perceived in the present, the possibility of the error in the perception – i.e. risk is always present. Uncertainty and risk are inherent in all investment activity, no matter what the size of the capital commitment, and we ignore it at our peril.

The risk dimension is a crucial factor in the appraisal of investments and it needs to be approached rationally. A rational approach to risk associated with investment must include: recognition and definition of risk and its various components; quantification and measurement of risk; the analysis of risk; and attempting to respond to risk.

An earlier chapter (Chapter 3) contains a comprehensive introduction to the principles of definition, classification, quantification and measurement of risk and its main components.

The purpose of this chapter is to examine more closely those analytical methods which promise to be most effective for the treatment of investment risk in general, and that would also be adaptable for the analysis of the risks associated with property investment assets.

It should be understood from the outset that the analytical methods examined in this chapter are expected to give assistance to the decision-maker by enhancing the appraisal of investment projects and not intended to cause more confusion in the processes of market valuations.

The methods of risk analysis must concentrate on the principal concern of the investor: are the risks associated with an investment adequately compensated by the expected returns generated? This criterion is often referred to as the **risk–return trade-off**. Furthermore, the examination of the risk–return characteristics of an investment is usually carried out relative to the characteristics of other investment assets – i.e. in the context of an existing or a planned investment portfolio.

The gap between the theory and practice of risk analysis is wide in all investment media but particularly in the medium of property. As with the primitive and implicit treatment of risk in the market valuation of property, the appraisal of property propositions also tend to concentrate on the 'qualitative' aspects of risk. The traditional approach to investment risk avoids the explicit treatment of risk and usually concentrates on a few simple assumptions about the future and relies very much on instinct and intuition. The complexity of the present-day investment scene, however, requires explicit, rather than implicit, risk analytical methods.

Little progress is discernible to date in the explicit, quantitative analysis of property investment risk. The arguments and debate about traditional methodologies resulted in delaying the evolution of risk analysis on similar lines as evolved for the risk analysis of stock market securities. Risk analysis, based on and evolved from modern financial theories, is now reasonably well established and gaining respect from practitioners. It is hoped that the analysis of property investment risk will make strides to catch up with the developments in other media.

Section 10.2 traces the evolution of the methods of risk analysis. In this section the fundamental concepts of the definition and measurement of risk will be re-established, and the main methods of analysis are outlined in Section 10.3. Section 10.4 examines the various **market-based approaches** to risk analysis, whilst Section 10.5 is entirely devoted to the use of the methods to analyse risk. Section 10.6 contains the overview of the rationale of the management of risk. Section 10.7 deals with the analytical methods evolved from **modern portfolio theory** focusing on the analysis of **market-related risk** and the use of **beta coefficients**.

10.2 FUNDAMENTAL CONCEPTS OF THE DEFINITION AND MEASUREMENT OF INVESTMENT RISK

10.2.1 Towards a suitable definition of investment risk

Investment is about future expectations. The investor's expectations about the returns from an investment are formulated in the light of the performance of the investment in the past, the security and stability of returns and, of course, the security of capital committed. Risk emerges as the expression of the predictability with regard to the future returns and the security of capital. The greater the dispersion of realized returns, then the precision of forecasting future returns will become the more difficult.

Investment and gambling have many similarities. Both require decisions about commitments now for uncertain rewards later. In making the decision either to place the bet or rejecting it the risk (i.e. the probability of reward or loss) will be weighed. The principal difference between an investment and a wager is that the rewards from the investment are expected after the elapse of a considerable amount of time, whilst a wager offers rewards or losses almost immediately. Another difference is that the rewards from an investment are dispersed in time in a series of instalments, whilst the winnings from a bet are usually available in one lump sum.

Great minds have been grappling with the attendant problems of investment and gambling over the past two centuries. The results of these labours are embodied in economic theory and the theory of games respectively.

The problem of investment is two-pronged: how to value the returns displaced in time, and how to translate risk into value. In modern investment evaluation procedures the value of time has been successfully resolved through discounting. The incorporation of time preference into the evaluation process is strictly valid only under conditions of certainty. Attempts to use the discounting process to deal with the problems of uncertainty led to a great deal of confusion. The risk needs to be considered separately and the methods of the theory of games appear to be promising to translate risk into value.

In gambling the sum of money required for a wager and the possible reward is known with certainty. All the gambler needs to do to accept or reject a wager is to weigh the chances of winning in the light of the commitment and the reward. The investor, on the other hand, has a perception of future uncertain rewards, and needs to know about the likelihood of achieving those rewards whilst take into account the time-value aspect.

A suitable combination of discounting and probability criteria is the best solution to the problem of rational decision-making. The way is now reasonably clear towards a suitable analytical definition of investment risk: *In the context of investment decision-making, risk is defined as the extent to which the actual outcome of a decision may diverge from the expected outcome. Investors regard the possible divergence from the expected outcome which result in lower return as most important.*

10.2.2 Analytical measures of risk

The measure of risk which would be most suitable for analysis must give a clear indication of the extent to which the actual outcome is likely to deviate from the expected value. The **mean absolute deviation** and the **standard**

deviation of a probability distribution are suitable in this respect. Although the standard deviation is a slightly more complex measure, it is preferable because of its mathematical superiority.

The standard deviation and its square the variance can be used as the absolute measure of variability of the actual outcome around the expected outcome; hence it can be regarded as the best albeit imperfect quantitative measure of risk.

The comparison between investment projects is not always satisfactory on the basis of *absolute* risk measures. The comparison would be more valid if *relative* risk measures were used; similarly, absolute risk measures are unsatisfactory when the variability is to be measured against a non-stationary benchmark or standard. This situation occurs when the variability of investment returns is to be examined relative to the market in which the investments are traded.

For the direct comparison of investment projects the simple statistical measure of the **coefficient of variation** is satisfactory, which is simply the ratio of the standard deviation of the outcomes and the expected value of the outcome:

$$\text{Coefficient of variation} = \frac{\text{Standard deviation of outcomes}}{\text{Expected value of outcome}}$$

The measurement of variability of investment returns in the context of the investment market and relative to the movements of the market require special consideration. In this case, variability will be related to the responsiveness of investment returns to market movements. To distinguish the relative measure of variability in the context of the market the term **volatility** is often used. The appropriate measure here is the **beta coefficient** of a particular investment, which is the expression of the statistical relationship between investment returns and the market and computed either by using a simple regression technique or by the formula:

$$\text{Beta} = \frac{\text{Covariance between asset returns and market returns}}{\text{Variance of market returns}}$$

All of the above measures of risk rest on the assumption that the investor could somehow perceive the distribution of the future rates of return. This is the point where most of the difficulties of the quantification and measurement of investment risk emerge. Risk and uncertainty are associated with events in the future and therefore to attempt to measure risk *objectively* is a contradiction in terms. On the other hand, the assumption that the past is a useful and reliable base on which a reasonably accurate view of the future could be formulated seems to be valid albeit in the short run. So by using the evidence from the past and the present prevailing trends, the investor should be able to perceive the shape of the distribution of future rates of return.

By using the rates of return figures achieved in the past, a **frequency distribution** can be drawn up and the appropriate statistics, the **mean, vari-**

ance and standard deviation could be computed. In the very short run, it could be assumed that the future will not be significantly different from the past and the shape of the frequency distribution will be very similar to the probability distribution of the future rates of return.

However, the precision of the above extrapolation will deteriorate very rapidly and the drawing up of probability distributions will rely on reasoned, subjective inputs. Although there are many problems associated with the use of subjective judgements and individual perceptions in the decision-making process, there are a number of reliable methods available to draw up probability distributions on a subjective basis; the discussion of these methods are outside the scope of this chapter. Here it is sufficient to say that with a little care and thought the beliefs of individuals or groups concerning future outcomes can be converted into very useful probability distributions.

10.3 AN OUTLINE OF THE METHODS OF RISK ANALYSIS

The analysis of investment risk is aimed at gaining insight into the effects of those factors which influence the rewards from the commitment of capital to investment projects. In investment decision-making the main concerns are the security of capital committed, the size and security of future rewards. Aspects of security (i.e. the risk dimension of investment) are best viewed and judged in terms of probabilities.

The risk dimension may be approached either by attempting to describe the riskiness of investments or by perceiving risk for a particular project and incorporating such perceptions into the appraisal model used.

Methods of analysis which attempt the description of the riskiness of investment projects include:

- expected value–variance method;
- sensitivity analysis;
- scenarios;
- simulation methods;
- analysis of Beta coefficients.

The methods, which attempt to incorporate subjective perceptions of riskiness by making 'appropriate' adjustments to selected components of the appraisal models are:

- risk-adjusted discount rate method;
- certainty equivalent method;
- sliced income method.

In practice, the above risk treatment methods are rarely used by themselves but in a combination of methods. By using a number of complementary analytical methods on the same investment project, the project's riskiness may be better understood and decision-making more effective. Some of the above methods are outlined below, others will be detailed in separate sections.

10.3.1 The mean–variance rule

When the expected returns and their riskiness, expressed by a measure of dispersion, has been perceived or calculated for a number of projects, an operational decision rule can be formulated. This decision rule is usually referred to as the **mean–variance rule or criterion**. This criterion simply states that if two investment projects A and B are compared, A will be preferred to B if at least one of the following situations arise:

(a) the expected return of A is greater than that of B and A's variance is less or equal than that of B's; or
(b) the expected return of A exceeds or is equal to that of B and A's variance is less than that of B's.

This criterion, however, cannot distinguish between projects in a portfolio setting where both the expected returns and the variances are different.

10.3.2 Risk and cash flow streams

Most investment appraisal models and decision models are cash flow based and use the DCF method to take into account the time value of money. In the presence of risk each of the cash flows will need to be specified according to the investor's perceived expectations. The cash flow expectations are expressed with appropriate probability distributions from which the expected value of a period cash flow and its variance may be computed.

The period cash flows may be totally independent of each other or they may be dependent on each other. The degree of interdependence of period cash flows may range from a slight dependence which may be ignored to total serial dependence. In this respect, the statistical term **correlation** is used to describe the state of affairs. The degree of interdependence or correlation of period cash flows determine the way the variance of the expected Net Present Value or the Internal Rate of Return of the cash flow stream is to be calculated.

The risk explicit form of the DCF model is as follows:

$$E(NPV) = C_0 + \frac{E(C_1)}{(1+r)} + \frac{E(C_2)}{(1+r)^2} + \ldots + \frac{E(C_n)}{(1+r)^n}$$

where $E(NPV)$ is the expected Net Present Value of the cash flow stream;
$\quad E(c_i)$ is the expected ith period cash flow;
$\quad C_0$ is the initial capital investment;
$\quad r$ is the risk-free rate of interest.

The risk associated with each period cash flow is measured by the variance or standard deviation of the perceived probability distribution of that particular cash flow.

When the risk measure (i.e. the variance of the NPV of the cash flow stream) is to be calculated, the knowledge of the degree of serial dependence or correlation of period cash flows is required. The calculations are reasonably simple if the period cash flows are totally independent or, in the other extreme, are perfectly correlated. In reality, most cash flow streams will be made up from period cash flows which exhibit some degree of dependence but not perfect correlation.

The variance of the NPV of totally independent period cash flows is computed as follows:

$$\delta^2(NPV) = \sum_{i=1}^{n} \frac{\delta_i^2}{(1+r)^{2i}}$$

The standard deviation of the NPV of perfectly correlated period cash flows is computed as follows:

$$\delta(NPV) = \sum_{i=1}^{n} \frac{\delta}{(1+r)^{i}}$$

The variance of the NPV of cash flow streams consisting of period cash-flows of some but not perfect correlation will be computed by taking the covariances of period cash flows into account. The numerical value of the variance of the NPV, in these cases, will fall somewhere between the higher variance of the NPV of perfectly correlated cash flows and the lower value of the variance of the NPV if the cash flows are totally independent.

Example:
Consider the investment requiring a capital outlay of £500 000 and with projected cash flows at the end of years 1 and 2 perceived in terms of the following probability distributions:

| | Cash flows | |
Probability	Year 1	Year 2
0.1	100 000	200 000
0.2	200 000	400 000
0.4	300 000	600 000
0.2	400 000	800 000
0.1	500 000	1 000 000

Compute the expected NPV of the project and measure the risk of the project if the period cash flows are: independent or perfectly correlated.

What are the chances that the project will produce a positive NPV if the risk-free rate of discount is 10%? Assume a near-normal probability distribution for the project's NPV.

Using the formulae shown above the expected period cash flows and their standard deviations are as follows:

$$E(C_1) = £300\,000 \qquad\qquad E(C_2) = £600\,000$$

$$\delta_1 = £109\,000 \qquad\qquad \delta_2 = £219\,000$$

from the DCF model:

$$E(NPV) = -£500\,000 + £300\,000/(1+0.1)^1 + £600\,000/(1+0.1)^2$$

$$= £268\,000$$

Assuming that the period cash flows are independent, risk associated with the project's NPV is as follows:

$$\delta(NPV) = \sqrt{\sum_{i=1}^{n} \frac{\delta_i^2}{(1+r)^{2i}}} = £206\,000$$

The standard deviation of the NPV if the period cash flows are perfectly correlated is computed as follows:

$$\delta(NPV) = \sum_{i=1}^{n} \frac{\delta}{(1+r)^i} = £280\,600$$

To attempt to answer the question: what are the chances that the project will produce a positive NPV if the risk-free rate of discount is 10%? assuming that the period cash flows are independent, we use the standard statistical procedure:

$$d = \frac{X_c - X}{\delta} = \frac{0 - £268\,000}{£206\,000} = -1.3$$

From the tables of normal distributions we find that the probability of the NPV being zero or less is 0.0968 or 9.68%. This means that the probability of a positive NPV is 90.32%.

10.3.3 Sensitivity analysis

The appraisal and analysis of most investment projects requires the examination of the effects of a large number of different factors. The effects of these factors are usually uncertain. The main problem is that conditions of uncertainty are such that neither are the estimates of the factors reliable in the form of single-point estimates nor is it possible to include the estimates in the form of probability distributions.

The identification of those factors which could have a major influence on the success or failure of the investment project is of crucial importance. By examining the change in the result of the investment as the consequence of the change in one of the factors, we may gain useful insights into the investment's behaviour under different circumstances and also into the influence of uncertainty. The process of the exploration of the change of the projected outcome of the investment resulting from the change in one of the factors of the project is **sensitivity analysis**.

The principle of sensitivity analysis is very simple and it is particularly useful to give an initial assessment of the potential impact of risk on the investment's profitability. Sensitivity analysis is not aimed at quantifying risk, but identifying the factors which are the potential sources of risk. The decision-maker is provided with answers to a number of 'what if...?' questions. Sensitivity analysis is widely used because of its simplicity and ease of interpretation.

Sensitivity analysis can be presented in a number of forms: tabulated results over a specified range of the selected critical variable, various sensitivity graphs which can provide answers at a glance, etc.

One of the most useful features of sensitivity analysis is that it can help to determine the 'break-even' values of the variables involved. Computerized investment appraisal procedures usually have extensive facilities for sensitivity analysis and the results are presented in sensitivity tables or in sensitivity graphs.

If the change of values of the critical factors are expressed in terms of the percentage deviation from the expected value, the graphs of the sensitivity relationships can be presented on a single graph. Figure 10.1 illustrates this form of presentation:

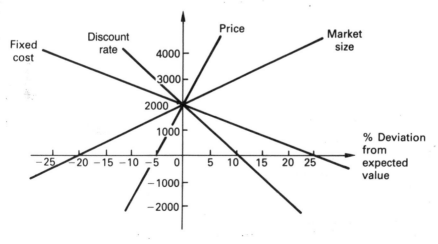

Fig. 10.1 Composite sensitivity graph.

10.3.4 Scenarios

The use of scenarios can improve the structure of sensitivity analyses and the expected value–variance method by grouping the estimates to suit particular combinations of circumstances or scenarios.

When using sensitivity analysis, the choice of the width of the range, over which the effects of the change of a particular factor is to be examined, may cause some difficulties. The simple device of grouping the appropriate values of all the factors into optimistic, realistic and pessimistic combinations can give a very useful insight into the overall range of possible outcomes.

If the groupings or scenarios are also considered in terms of the likelihood of their occurrence, then an expected value for the outcome of the investment can be crudely formulated.

The improvements provided by the use of scenarios is slight because the probability assessment of the occurrence of the scenarios envisaged is either absent or very crude. The example below attempts to illustrate the presentation of results using the scenario approach:

Scenario	Investment worth	IRR	Growth estimate
Optimistic	£580 000	13%	9.5% p.a.
Realistic	£520 000	12%	8.0% p.a.
Pessimistic	£460 000	10%	7.5% p.a.

If the probabilities of occurrence of the scenarios is also considered, the results may appear as follows:

Scenario	Probability	Investment worth	IRR	Growth estimate
Optimistic	0.2	£580 000	13%	9.5% p.a.
Realistic	0.7	£520 000	12%	8.0% p.a.
Pessimistic	0.1	£460 000	10%	7.5% p.a.

$E(\text{Investment worth}) = 0.2(580\,000) + 0.7(520\,000) + 0.1(460\,000) = £526\,000$

10.3.5 The investor's attitude to risk

In the making of decisions about investments whose returns are uncertain the investor's attitude to risk will play an important part. Investors usually prefer certainty to uncertainty and less risk to more risk, whilst the promise of higher returns will always be regarded as more attractive than lower returns.

In fact even unsophisticated investors will trade risk for return. The ranking of investment alternatives with differing risk–return characteristics is basically a psychological process which is very difficult to unravel as it varies from one individual to another. It would be nice and convenient if these processes could be rationalized and the rules governing them defined for others to follow. Unfortunately, these processes can be modelled only in very basic and primitive terms, always assuming that the investor will always react rationally in the face of risk and uncertainty. Decision-making could be rationalized if a model would be available which describes adequately the investor's perceptions of, and reactions to, investment risks.

The model representing the investor's attitude towards risk is usually constructed by using the concept of utility. Utility replaces monetary rewards as the expression of satisfaction in risky situations.

An individual's utility function can be constructed in a number of different ways. The simplest approach is to ask the individual to state the price he would be willing to pay for an investment which produces two different pay-offs but with the probabilities of the occurrence of the pay-offs varying over a range. The individual's responses are noted and presented in graphical form; typical shapes of utility functions are shown in Fig. 10.2.

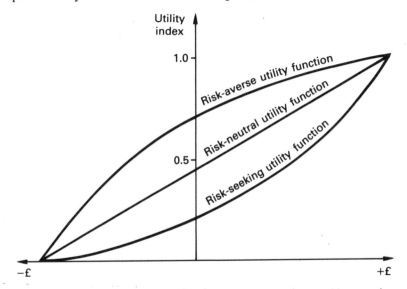

Fig. 10.2 Typical utility functions.

10.4 MARKET APPROACH TO RISK ANALYSIS

The traditional or market approach to the treatment and analysis of risks associated with property investment projects uses the framework of the income method of valuation. In the past decade, however, DCF appraisal models have gained ground. The presence of uncertainty and risk is taken into account in an implicit way by the use of the **all risks yield** (ARY) and suitably adjusted discount rates. All adjustments are intuitive, although they are supposed to be based on the analysis of comparables available in the market. Risk analysis in this setting is nothing more than the examination of the effects of variations of the main input factors over a perceived range of values. The rational and systematic analysis of investment risk is usually avoided and crude adjustments for perceived risks are the order of the day.

The use of single-point estimates of future income levels and cash flows is most common, together with the all risks yield and with risk-adjusted discount rates.

The principal sources of risk and uncertainty are the tenant's covenant, rental levels after reviews, vacancy and the costs of maintenance and management. In recent years other sources of risks, such as the level of liquidity of capital, obsolescence and depreciation, have also been considered. The effects of all these risk factors are to be reflected in the all risks yield and the discount rate used in the DCF analysis.

Where a particular investment is subject to exceptionally great uncertainty, the analysis of risk takes the form of simple sensitivity analysis. Although sensitivity analysis provides some measure of the possible dispersion of the expected results, it offers no indication as to the relative likelihood of one adjusted value being more precise than any other. The main advantage of the sensitivity approach is that it highlights those risky variables which require particular attention in the calculations.

A further variation of the sensitivity analysis approach is the use of scenarios. In this approach, estimates of the input variables are grouped together to reflect optimistic, realistic and pessimistic situations or scenarios. The results are expected somewhere between the optimistic and pessimistic scenarios.

Since those decision-makers who are reluctant to separate investment risk from its market environment generally prefer single-point estimates of the expected results of their investment activities and find the range of results provided by sensitivity analyses or by other analytical techniques unhelpful, they demand appraisals which give the conclusions in single-point form but also take account of the presence of risk.

The following risk adjustment techniques have been suggested to improve the market-based treatment of risk and to satisfy the requirements of the traditional market-oriented investor:

1. Risk-adjusted discount rate (RADR).
2. Certainty equivalent technique using the risk-free discount rate (RFDR).
3. The sliced-income approach.

10.4.1 Risk-adjusted discount rate (RADR)

All the adjustments to the **all risk yield** or the DCF discount rate are based on a perceived required rate of return which compensates for all the investment risks. This required rate is based on the risk-free rate of return, plus a subjectively determined premium which is expected to compensate the investor for the extra risks involved.

The selection of the risk-free rate of return is relatively simple in practice. In most cases, the yield on government stock, either undated or dated to match the life of the investment under scrutiny, is adopted as the riskless rate.

The meaning of the riskless rate of discount needs further explanation. In the context of investment analysis, risk-free does not imply the total absence of risk, but the virtual absence of the risk of default. Some other risks remain to be reckoned with: mainly inflation risk and the adverse changes in the structure of interest rates.

The real problem is the selection of the premium which must be sufficient to compensate for the additional risks associated with the investment under consideration, over and above the returns guaranteed from investments in risk-free securities. Theoretically this should be derived from the investor's risk–return trade-off function. This function is rarely, if ever, available. In the absence of a precise risk–return trade-off function, the risk premium is likely to be set with a great degree of subjectivity.

The required rate of return or risk-adjusted discount rate is constructed as follows:

$$RADR = (1 + i)\,(1 + g)\,(1 + r) - 1$$

where $RADR$ is the adjusted interest or discount rate, and:

i represents the value of time or time preference;

g is the compensation for inflation;

r the reward for risk bearing.

In practice, risk-adjusted discount rate is constructed in a simpler but very much cruder way:

Required rate of return = Risk-free rate of return + Risk premium

or

RADR = RFR + RP

The main flaw of the RADR is the subjectivity of the selection of the risk premium. The other main drawbacks are the increasing discount applied to future cash flows which carry the danger of double discounting and that risk premia have to be set for each individual project.

In practice, this method of accounting for risk tends to be unreliable; however, because of its relative simplicity, it remains popular for the initial screening of investment propositions.

10.4.2 Certainty equivalent technique

It is thought that some of the difficulties associated with the use of the RADR may be overcome by first adjusting the projected cash flows of the investment to cash flows which are achievable with a reasonable and calculable degree of certainty. The cash flow stream, converted into a stream of certainty equivalent cash flows, is then discounted at the riskless rate of discount (RFR or RFDR).

The main problem of this method is the determination of the certainty equivalent cash flows. This may be done by using 'best estimates' or by using the standard deviation of the perceived normal distribution of the expected cash flows. The latter method appears to be the more reasonable of the two, but more complicated in practice.

10.4.3 The sliced-income approach

This technique emerged quite recently and is aimed to improve the above two methods. In essence, this method is a DCF model using the hardcore approach with the assumption that the core income is guaranteed and therefore it should be discounted at a risk-free rate. The additional incomes expected after rent reviews or reversion are to be discounted at the risk-adjusted rate to reflect their more risky top-slice nature. However, the problem of the choice of the risk-adjusted rate remains, although its significance diminishes when the new income slice is relatively thin.

All the above methods are loaded with problems and difficulties. Nevertheless, they are attempting to account for risk explicitly without providing many facilities for analysis. They are particularly suitable for the initial screening of new property investment propositions. For the provision of proper analytical facilities the investor must turn to probabilistic simulation methods.

10.5 USING THE METHODS TO ANALYSE RISK

10.5.1 Traditional methods of taking risk into account

There are several ways in which traditional appraisal methods attempt to cope with risk and uncertainty. It was generally accepted that risk and uncertainty were functions of time, although the relationship between risk and time has never been totally explored or demonstrated. The riskiness of projects were perceived in the light of the length of time required to repay the capital invested. The **payback method** of investment appraisal has been popular for a long time as it incorporates the above-mentioned view of risk. The longer the payback period, the more risky an investment project is considered to be. The main objection to this particular method of assessing risk is that it takes no account of the time value of money. Another traditional approach to risk is to consider investment projects within a limited time horizon.

Sensitivity analysis has been used quite extensively to give insights into the risk characteristics of projects which were appraised by using standard, deterministic appraisal methods. Although the sensitivity analysis approach will not produce a single risk measure for a particular investment project, nevertheless it has been proven to be a very useful tool in decision-making. Sensitivity analysis is often the only means to assess the risk aspect and it is frequently used to underpin decisions made, employing more advanced and sophisticated risk assessment methods.

Another early method to account for the presence of uncertainty is to use 'most likely' values in the computations. These values were modal or average values deducted from some samples of the input variables used. The samples were either drawn from historical evidence or were a collection of forecasts of experts. Such values when used in the calculations were assumed to produce a result which was 'most likely' to occur. This approach is often extended to use the known extreme values of the input variables to produce a so-called **pessimistic** and **optimistic estimate**, together with the most-likely of the outcomes of investment projects. This method is unsatisfactory, partly because it dismisses the possible variations of the outcomes around the estimated values and partly because it is most unlikely that all the pessimistic and all the optimistic forecasts will ever come together. In fact these optimistic and pessimistic estimates of the outcomes are unlikely to ever occur in reality.

10.5.2 Some simple approaches to the treatment of risk

The expected value or scenario approach

The basis of this method is that it is possible to produce a range of cash flow profiles for a particular investment project where each individual estimate of cash flow profile would be perceived in a particular scenario or state of the world. For example, a cash flow profile could be drawn up under expected boom conditions, another profile could be drawn up for a state of economic depression, whilst yet another profile could be envisaged under normal economic circumstances. For each cash flow profile an internal rate of return can be calculated. If the likelihood or probability of the alternative scenarios is combined with the respective internal rate of returns produced, the mean of such internal rate of returns will be the expected internal rate of return of the project. Whilst this type of approach to the appraisal, under conditions of uncertainty, is useful, it cannot be considered to be an assessment of risk as the expected value will not have any indication of how far the actual performance may diverge from what is expected. The following example illustrates this procedure.

An investment project is expected to generate the following cash flow streams under boom, normal and depressed state of the economy. The probabilities of the occurrence of the various states of the economy are also assessed as shown:

Scenario	Boom	Normal	Depression
Probability	0.2	0.6	0.2
Period		Cash flows	
0	−10000	−10000	−10000
1	5500	5000	4000
2	6000	5500	4000
3	6400	6000	4000

Rates of return: $IRR_B = 37.8$ $IRR_N = 28.8$ $IRR_D = 9.7$

Expected IRR $= (0.2)(37.8) + (0.6)(28.8) + (0.2)(9.7) = 26.8\%$

The risk-adjusted discount rate method

This method is based on the assumption that in an uncertain world the market-determined interest rates are a combination of two separate components: the marginal rate of time preference which is often called the risk-free or default-free interest rate, and a risk premium which represents the market's estimated reward for risk taking. Usually this can be expressed as follows:

$$R_x = r_f + r_p$$

where R_x is the expected return adjusted for risk on an investment project x;

r_f is the risk-free rate, say, the current rate of return, on some long-term government bonds;

r_p is the risk premium required for the additional risk born on this particular investment project x.

The risk-adjusted discount rate approach is useful to the decision-maker as it does produce decision advice in the form of a risk-adjusted target rate of return. The main disadvantage is that the actual risk premium is still a subjective measure and hence the reliability of the method is often suspect. Furthermore, the application of risk-adjusted discount rate involves an often undesirable side-effect, as through the compounding process, the effect of the risk premium increases as the life of the project increases.

The downside risk method

This method is similar to the risk-adjusted discount rate method; however, it utilizes some of the properties of normal probability distributions. It estimates the likelihood of the probability of the project not achieving a particular target rate set. Such a target rate, at the first instance, could be the risk-free rate. Using the properties of the normal distribution curve, the probability of the project not achieving the risk-free rate of return could be estimated. If, then, risk-adjusted discount rate is perceived, the procedure could be repeated to compute the likelihood of the rate of return falling below the risk-adjusted target rate. In the light of these probabilities, the appropriate decisions could be made. The following example illustrates the computational procedures:

Considering the same investment project used in the previous example, the following additional information is available.

The current risk-free rate of return, say, index-linked gilts, is 8% and the project is required to produce an additional risk premium of 3%. It is assumed that future rates of return will be normally distributed around the expected rate of return.

Scenario	Boom	Normal	Depression
Probability	0.2	0.6	0.2

Period		Cash flows	
0	−10000	−10000	−10000
1	5500	5000	4000
2	6000	5500	4000
3	6400	6000	4000

Rates of return $IRR_B = 37.8$ $IRR_N = 28.8$ $IRR_D = 9.7$

The Expected IRR $= (0.2)(37.8) + (0.6)(28.8) + (0.2)(9.7) = 26.8\%$

The variance of returns $= \Sigma(IRR_x - \text{Expected IRR})^2.(\text{prob}_x)$

$$= (37.8 - 26.8)^2 (0.2) + (28.8 - 26.8)^2 (0.6)$$

$$+ (9.7 - 26.8)^2 (0.2)$$

$$= 85.082$$

The standard deviation of returns $= 85.082 = 9.224$

Using the tables (Appendix) to find the area under the normal curve representing the probability of *not achieving*, the

Risk-free rate of 8% $= \dfrac{8 - 26.8}{9.224} = 2.038 \quad 0.02068$

Risk-adjusted target rate of 11% $= \dfrac{11 - 26.8}{9.224} = 1.713 \quad 0.0436$

Therefore the probability of the project *not achieving*

8% risk-free rate of return is 2.1%

11% risk-adjusted rate of return is 4.4%

10.5.3 Some more sophisticated approaches to the treatment of risk

The second group of methods for the treatment of investment risk is associated with portfolio theory. The essence of these methods is the realization that investment risk can be reduced by creating a portfolio of investment assets. The risk structure of investment portfolios depend on the degree of correlation between the returns produced by the component assets of the portfolio and the differences between the risk characteristics of the individual assets.

The unsystematic component of the total risk of the portfolio can be diminished through diversification.

These methods demonstrated the validity of the use of the **variance** as an appropriate quantitative measure of risk, and **volatility** expressed by the beta coefficients as an indicator of the systematic component of total risk.

The second group of methods provide means to analyse and account for risk by using **simulation procedures**.

Simulation is one of the most powerful analytical tools available for the decision-maker under the conditions of uncertainty and risk. Most of the decision problems associated with investment involve a careful analysis of complex problems, as most investment projects are subject to the influence of a number of variable factors. If a model representing the behaviour of a particular investment proposition or project could be created, then the decision-maker could subject this model to the influence of the various and variable factors, and from such an experiment he could glean the likely behaviour pattern of the investment project. The model used should clearly reflect the understanding and the effects of the factors influencing the project. The reliability of the conclusions drawn from the experiments is entirely dependent on the precision of the model describing reality as closely as possible.

Simulation is a numerical process involving mathematical models. There are two main approaches to simulation, one using the methods of **game theory**, whilst the other uses a repetitive process of experimentation. When game theory is used, the simulations are characterized by some form of conflict of interests between people within the framework of a simulated environment. Business games are typical examples of this approach.

The other type of simulation, using repeated experimentation, can be divided into two types:

1. Using deterministic simulation models. In these kinds of model, the variables are assumed to have certain numerical values. These models will attempt to answer 'what if?...' type questions; and when the analysis is repeated, the process is called sensitivity analysis.
2. Using probabilistic (or stochastic) simulation models. In these models, the variables can be either assumed to be known with certainty (i.e. they are regarded as constants) or a number of the variables are represented not by known single values, but by their probability distributions. Those variables which are represented by their probability distributions are regarded as random variables and are outside the control of the decision-maker. They are also regarded as the variables representing risk and uncertainty and their aggregation in the model will bring about the behaviour pattern of the model itself.

The **Monte Carlo method** is a simulation technique for the analysis of the behaviour of probabilistic simulation models. It is particularly useful in dealing with models where some of the variables are of the stochastic type.

The aim of the Monte Carlo simulation is to allow the experimentation with the model, so that it can behave realistically and exhibit the characteristics of the real thing, with reasonable precision. By allowing the various components of the model to interact in a realistic, random fashion by observing the behaviour of the model, reasonable conclusions may be drawn.

The name Monte Carlo derives from the fact that sampling from probabilistic distributions is an essential part of carrying out such simulation. In making a Monte Carlo simulation run, the values of the uncertain variables are obtained by random sampling from their respective distributions, using random numbers generated in an appropriate way. These values are then aggregated in the model, according to the logic of the model, and a result is produced. The result thus obtained is then stored. This procedure is repeated and the results stored away until a statistically significant sample of results is stored. The sample of stored results is then analysed and its distribution described with appropriate statistics. The distribution of the sample and its statistics is considered to be the reflection of the expected behaviour of the model in reality.

The Monte Carlo risk simulation involves the following steps:

(i) Constructing a model which could describe the behaviour of an investment project over a period of time. Most frequently, the discounted cash flow models are used in either the NPV or IRR modes. For an investment project, these DCF models will give a behaviour profile reflected in their perceived cash flow streams. These cash flow streams will be expected to be subject to the influences of various factors in different points in time, in a random fashion.

(ii) Determining the certain and uncertain variables with their constant values and their probability distributions respectively. The drawing up of the expected probability distributions of the probabilistic variables is either done on a objective, or more often, on a subjective basis.

(iii) Completing a number of simulation cycles using random number based selection for the uncertain variables and computing and storing results from the simulation model adopted; this process is shown in Fig. 10.3.

(iv) Computing the statistics for the sample of results generated by the simulation cycles and analysing and interpreting them.

The careful interpretation of the results produced by the Monte Carlo method is absolutely essential as they could be potentially misleading. A particularly disturbing aspect is that the failure of the recognition of the interdependencies of the variables can lead to a completely distorted picture of the expected behaviour of the model. Furthermore, there is the question of the number of simulation cycles required to reach reasonable conclusions. It is generally accepted, in practice, that a few hundred simulation cycles would produce a sufficiently reliable reflection of the behaviour of a model. Theoretically the number of simulations required could be calculated from an appropriate mathematical formula.

In practice, the main drawback of this method is the difficult and rather haphazard nature of the specification of the probability distributions of the uncertain, risky variables and the danger of ignoring the interdependencies existing between them.

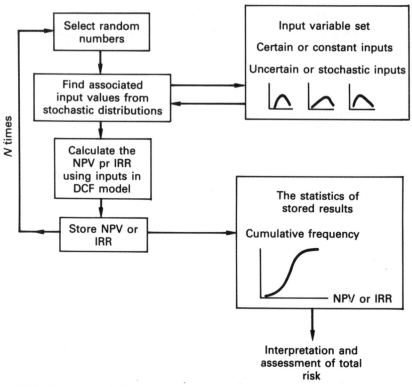

Fig. 10.3 The Monte Carlo process.

The large number of computations involved in the repeated simulation cycles necessitate the use of computers. The computer programs used should have facilities to incorporate various models best suited to describe the behaviour of individual investment projects.

It should be clearly understood that the Monte Carlo simulation process can never produce a precise, well-defined conclusion. The main virtue of this method is that it explores the structure of the problem and apportions due weight to the effects of the individual, uncontrollable random factors. For these reasons the variability of the expected behaviour (i.e. the riskiness) of a project can be observed and assessed.

10.6 THE MANAGEMENT OF RISK

It is generally accepted that rational investors aim to maximize future returns whilst striving, at the same time, to minimize the investment risks involved. The main problem of managing the risk aspect of investment is the lack of reliable information about the probability of various events occurring in the future which could adversely affect investment projects. Decisions need to be made in anticipation of such uncertain events. Unfortunately, apparently good decisions at a particular point in time can produce bad results in the future. There is no way to guarantee the final outcome of decisions made. It should be reasonable to judge the quality of a decision with regard to its likely consequences on the basis of the information at the time rather than in hindsight. Good decisions in the face of risk are the ones that tend to minimize the prospect of undesirable consequences. Undesirable events in the future will inevitably involve some costs, therefore the expected costs of such events could be computed by multiplying the probability of the event by the actual cost if it does occur. As a consequence, if the probability of the event could be reduced, the associated expected costs could also be reduced. The effective management of investment risks require:

(i)   to consider the likelihood and impact of such events occuring in the future;
(ii) to devise appropriate strategies for controlling such uncertainties and to minimize their impact;
(iii) to incorporate such strategies into the general decision-making framework.

A rational investor who treats investment risk prudently should:

(i)   formulate his investment objectives precisely with regard to return and risks;
(ii) strive to identify all possible risks, particularly the major ones;
(iii) eliminate certain risks, transfer others and attempt to minimize the remaining risks;
(iv) decide whether or not the expected returns are worth the risk still remaining.

One of the most effective ways of reducing investment risk is through **diversification**. By diversifying, the investor could virtually eliminate the unsystematic component of the total investment risk of his investment portfolio. The remaining systematic risk could be effectively managed by carefully selecting only those assets for the investment portfolio which will keep the volatility of the portfolio at an acceptable level. The investor is also able to control the total risk of his investment portfolio by investing in risky and risk-free assets in appropriate proportions.

As far as the risk management of individual investment projects are concerned, effective strategies could be formulated by using the Monte Carlo simulation process to outline the risk characteristics of the project.

10.6.1 Monte Carlo simulation and property investment risks

The Monte Carlo technique, at first sight, appears very capable to provide insights into risk aspect of property investment projects. This is certainly true, provided that the operational model of the investment is properly constructed, the nature and interrelationships between the input variables have been identified and taken into account and, finally, the number of simulation runs will produce a statistically significant sample of results.

The base model is usually a DCF model configured either for the calculation of the project's NPV or its IRR. The period cash flows are usually of the type generated from a preceding period cash flow by applying appropriate growth rates, depreciation, etc. Models based on the hardcore or term and reversion methods can also be used.

The risky variables can be easily identified but their estimation in terms of probability distributions and their degree of interdependence cause serious problems. The specification of the probability distributions of the risky variables can be improved with practice but the specification of the covariances needed to incorporate dependencies into the analysis is notoriously difficult. However, if the dependencies are ignored, the reliability of the analysis will virtually disappear.

The question of the number of simulation runs needed has been side-stepped in the past on the excuse that the number of simulations easily and cheaply available on computers will be more than enough to generate a statistically significant sample of results.

It should be clearly understood, that the Monte Carlo simulation process can never produce a precise, well-defined conclusion.

The main virtue of this method is that it explores the structure of the problem and apportions due weight to the effects of the individual, uncontrollable random factors. For these reasons the variability of the expected behaviour (i.e. the riskiness) of a project can be observed and assessed.

The Monte Carlo risk analysis of a typical freehold investment proposition would require the following inputs:

Base model: DCF IRR model
Holding period: 10 years
Capital investment: £500 000
Current rental income: £12 000
Current rental value: £24 000
Rent reviews: next review in two years' time and every two years thereafter, upwards only

Risky variable	Range %	Probability %
Annual rate of growth	2	10
	3	20
	4	50
	5	10
	6	10
Capitalization rate on disposal	4	5
	4.5	10
	5	20
	5.5	40
	6	15
	6.6	10

10.7 MARKET-RELATED RISK: THE ANALYSIS OF BETAS

Relatively recent developments in financial theory caused a major shift in the perceptions about investment risk. Investors realized well before the theoreticians that diversification, which was hailed as the panacea in the minimization of investment risk, had serious limitations. Diversification could only help in eliminating only a part of the total investment risk. The realization that the total investment risk can be partitioned into two main components initiated a completely new approach to the analysis and treatment of risk. The two main components of investment risk were named **systematic risk** and **specific or unsystematic risk**. These risk components were different in nature and required different analytical treatment.

The specific risk component, as its name suggests, is the product of the unique characteristics of an asset. This risk component can be virtually eliminated in a portfolio setting through diversification. Since the specific risk can be neutralized, the pricing process of the investment market will not allocate any premium for bearing this kind of risk.

The other risk component, the systematic risk, is to do with the market in which the investment asset is traded and it is related to the movement of that market. This is the reason why the systematic risk is also referred to as **market risk**. Market risk is non-diversifiable and associated with the general movements of the economy, and it affects all market-quoted assets to greater or lesser extent.

The 'beta coefficient', β, expresses the responsiveness or Volatility of the expected asset returns to the changes in the market. The beta coefficient is also regarded as a measure of the systematic risk of an asset or security.

If the numerical value of β_x is equal to 1.0, then the asset returns will move in unison with the changes in the market. If β_x is less than 1.0, then the asset returns will exhibit some sluggishness in following the changes in the market, whilst β_x values greater than 1.0 will mean an increase in volatility. Figure 10.4 illustrates how the beta coefficient represents market risk.

To calculate the systematic or market risk in terms of the absolute measures of variability the figure representing the risk of the market is needed. The variability or risk of the market is usually represented by the variability of a market index, say, one of the FT market indices:

$$\begin{matrix} \text{Market risk of a} \\ \text{security} \end{matrix} = \begin{matrix} \text{Beta coefficient of the} \\ \text{security} \end{matrix} \times \begin{matrix} \text{Variance of a} \\ \text{market index} \end{matrix}$$

or:

$$v_{jm} = (\beta_j) \times (V_m)$$

10.7.1 The analysis of betas

Understanding, measuring and analysing market risk are fundamental to investment decisions in the complex modern investment markets. Basically,

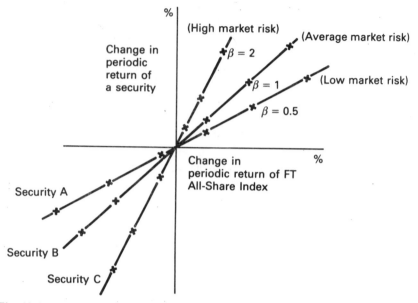

Fig. 10.4 Market risk as represented by the beta coefficient.

understanding market risk is understanding price behaviour. Market risk is the likelihood of the loss of capital or the reduction of returns owing to the changes in prices in the stock or other investment markets.

As an example, let us examine the risks associated with a stock whose relationship with the market is shown in Fig. 10.5. The stock has a beta coefficient of 0.99. This indicates that this stock has low market risk, and as far as the past is concerned the stock appears to have moved with the market. The alpha coefficient is also slight, 0.01, indicating a low unsystematic risk component.

The causes of price changes are beyond the control of the participants in the markets. Prices are influenced by psychological factors. Since the random fluctuations of the markets cannot be predicted, the investor must exploit all information available about the trends prevailing in the market.

There are number of ways in which investors can approach the problem of market risk.

The first step is to examine the historic behaviour of the price of the asset or security for any cyclical pattern or longer-term price and growth movements. In this examination of past records the market risk indicator, the beta coefficient of the asset or security, will be highly important, bearing in mind that the greater is beta, the greater the market risk. Although the beta coefficient is not constant, it alters slowly in time. Historic beta values may need to be adjusted in the light of the emerging trends and perceptions about future conditions when the assessment of the future market risk levels are necessary.

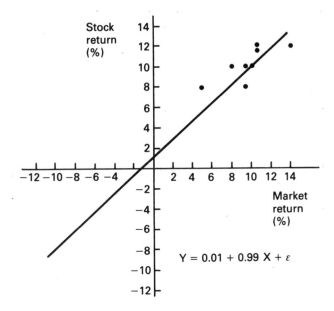

Fig. 10.5 The beta and alpha coefficients of a stock.

From the examination of past behaviour of some selected assets the respective market risk characteristics may emerge, and the investor may discard those assets with the large market risk content or those which are about to enter the downward slope of their price cycle. The timing of purchases or disposals can be decided from this analysis.

There are a number of additional difficulties arising from the use of beta coefficients to express the market risk associated with property assets. First of all, in the property investment market the term 'market risk' usually means **illiquidity** and not volatility as in the stock markets. Hopefully, this confusion of terminology will soon disappear. The main problem with property betas is that the property market is a fragmented, highly inefficient market and in such an environment beta coefficients are unreliable measures of the elusive market risk. Although property indices are becoming respectable, it will take a long time until they can be put on a par with their stock market counterparts.

10.8 SUMMARY

The past three decades have witnessed a revolution of increasingly advanced and sophisticated techniques in investment appraisal. However, the inclusion of the analysis and monitoring of the impact of risk and uncertainty on investment decisions are still relatively undeveloped and unsophisticated. The importance of risk has been realized long ago, although there are still serious arguments about a proper quantitative definition of risk. There are a variety of risk measures available to the investor, none of them perfect, and there is no standard accepted measure of risk. It has been generally accepted that risk should be regarded as describing the extent to which the actual outcome of a decision may diverge from what is expected, thus the greater possible divergence, the greater is the degree of risk. To a rational investor who is adverse to risk, the most important aspect is the possible divergence from the expected outcome of an investment decision which result in a lower return – i.e. he is particularly interested in the downside risk.

The definition, quantification, measurement and effective treatment of risk represent the most challenging aspect of investment decision-making. The effects of risk and uncertainty cannot be avoided in any investment activity. Investors are well aware of the dangers of trying to ignore these possible perils, although they do tend to get frustrated with the slow progress in the development of a suitable and effective methodology.

The definition of investment risk and its components is still subject to debate. There are a number of verbal definitions, a number of which are unsuitable for serious analytical work, and defying all attempts of quantification and measurement.

It is now generally accepted that uncertainty and unpredictability are one and the same. As unpredictability is usually associated with variability, the consensus is to define and measure risk and uncertainty with some measure of variability. There are a variety of measures of variability available for this purpose, each having some attractive and some awkward features. However, the variance and standard deviation seem to emerge as the most popular, absolute measures of variability, most suitable to measure risk.

A highly important development was the partitioning of the **total investment risk** into the **systematic or market-related risk**, and again into the **unsystematic or specific risk**. The systematic risk required an appropriate, relative measure of variability: volatility. This development has also had a profound effect on the treatment and management of risk.

A number of different approaches have also emerged for taking risk into account when investment decisions are to be made or when the monitoring of investment performance and efficiency are the main objectives. There is a promising progression in the development of such approaches. Although the traditional early approaches are still favoured by many, the more sophisticated probabilistic approaches are now gaining ground. With the advent of abundant and relatively cheap computer power, the simulation of risky situations will give the much desired means to assess risk.

FURTHER READING

Books

Lumby, S. (1981) *Investment Appraisal and Related Decisions*, Nelson, London.

Darlow, C. (ed.) (1982) *Valuation and investment appraisal, Estates Gazette*, London.

Merrett, A.J. and Sykes, A. (1974) *The Finance and Analysis of Capital Projects*, 2nd edn, Longman, London.

Byrne, P. and Cadman, D. (1984) *Risk, Uncertainty and Decision-making in Property Development*, E. and F.N. Spon, London.

Hull, J.C. (1980) *The Evaluation of Risk in Business Investment*, Pergamon, Oxford.

Coyle, R.R. (1972) *Decision Analysis*, Nelson, London.

Sharpe, W.F. (1985) *Investments*, Prentice-Hall, Englewood Cliffs, N.J.

Markowitz, H. (1959) *Portfolio Selection: Efficient Diversification of Investments*, Wiley, New York and London.

Bank Administration Institute (1968) *Measuring the Investment Performance of Pension Funds for the Purpose of Inter-fund Comparisons*, BAI, Park Ridge, Illinois, USA.

Articles

Hertz, D.B. (1964) Risk analysis in capital investment, *Harward Business Review*, January–February, 95–106.

Robicheck, A. (1975) Interpreting the results of risk analysis, *Journal of Finance*, December.

Sykes, S.G. (1983) The assessment of property investment risk, *Journal of Valuation*, **1** (3), 253–67.

Robinson, J. (1987) Cash flows and risk analysis, *Journal of Valuation*, **2** (5), 268–89.

Baum, A. (1987) Risk-explicit appraisal: a sliced income approach, *Journal of Valuation*, **2**(5), 250–67.

Hillier, F. (1963) The derivation of probabilistic information for the evaluation of risky investments, *Management Science*, **9**, 443–57.

Young, M. (1977) Evaluating the risk of investment real estate, *Real Estate Appraiser*.

Frazer, W. (1985) The risk of property to the institutional investor, *Journal of Valuation*, **1** (4), 45–59.

The measurement and assessment of investment performance

PART THREE

The rationale of the appraisal of investment performance

11.1 INTRODUCTION

Performance is defined as the achievement relative to some objectives. If the achievement can be measured, in quantitative terms, then the degree of achievement of the objectives set must also be expressed in a quantitative form.

Investment performance, and its measurement and analysis, have attracted considerable attention, particularly in recent years. A whole industry has developed offering professional advice concerning the measurement of portfolio performance and the interpretation of the results of the measurement.

Most of the activity and development have so far been limited to the performance measurements of investment portfolios containing stock market securities. Although property assets represent a substantial portion of the global institutional investment portfolio, it has been considered too complex to include them in the overall performance measurement of funds. Property assets and property portfolios have been treated separately from the point of view of investment performance.

The ultimate objective of an investment manager is to construct and maintain successful investment portfolios. It is absolutely essential therefore that appropriate methods are made available by which the success of investment assets and portfolios can be assessed. It is only through the continuous analysis of the achievements that appropriate decisions can be made to improve the efficiency of investment activity.

Earlier performance appraisals have tended to be rather anecdotal and descriptive. Very little analysis was attempted and only few explanations were hazarded to explain the results. There are inherent conceptual and technical problems in the measurement of performance and its explanation is also fraught with difficulties.

The most important development in large-scale investment activity in recent years has been the growth of performance measurement and portfolio analysis services. Two decades ago performance measurement was mainly concerned with the relative performance of investment funds, providing overall return on funds in order to provide the comparisons.

Performance measurement has now become highly sophisticated, providing detailed in-depth analysis of whole funds, portfolios and even individual investment assets. The analysis of performance extends beyond the analysis of performance of the investment portfolios and individual assets to the appraisal of the quality of management of funds and portfolios. The demand for such sophistication comes mainly from the institutional investor but increasingly fund managers are becoming interested not only in their own performance, but the management performance of their subordinates.

In the 1980s all investors, particularly institutional investors, became acutely aware of the need for the reliable measurement of the performance of their property assets. Whilst in the 1970s the inflation hedge provided by property investments obviated the awkward questions about investment performance. Investment in property could be justified on the basis of the provision of some protection against inflation and also on the basis of the need for diversification. More recently, however, the flow of money into property investments needs to be justified on the basis of performance prospects.

Property is capable of producing large and stable returns in the long run, although its short-term performance is subject to alarming fluctuations. Through the monitoring and analysing of property portfolio performance, the investor can gain valuable insights into the investment characteristics and behaviour of property assets and can find satisfactory explanations for the behaviour of property portfolios in the context of the movement of both the property and the general investment market.

Only through the measurement of performance can the true contributions – good or bad – of property assets with a global investment portfolio be evaluated. The special investment characteristics of property assets and portfolios require expert specialist treatment.

Attempts to transplant the methodology, used successfully for the measurement of investments of stock market security portfolios, into the measurement of the performance of property portfolios and assets have long been frustrated. The appropriate measurement methodology is still subject to evolution.

The rationale of the performance measurement and analysis of property assets and portfolios is the same as for other investment assets and investment portfolios. The main thrust of this chapter, however, is the examination of the rationale of the measurement and analysis of the performance of property investment portfolios.

11.2 PURPOSE AND OBJECTIVES

By measuring the performance of investment portfolio, the degree of achievement of a set of objectives and targets can be expressed in quantitative terms. The short fall or excess relative to the targets can then be analysed and useful conclusions and explanations drawn for strategic and operational decision-making.

It is very important to draw a clear distinction between the measurement of investment performance and the analysis of investment performance. Measurement without subsequent analysis is virtually useless.

Performance measurement and analysis may be used as a tool to improve the future performance of the portfolio, either by helping the investor to allocate the available funds in a most effective manner amongst various portfolios or by providing the individual portfolio managers with useful guidance and thus helping them to improve their future performance.

Performance appraisal (i.e. measurement and analysis) is a vital component of the investment decision-making process. Rational decision-making, at all levels, would be virtually impossible without the quantified evidence of past performance and a reasoned assessment of likely future performance. Performance appraisal is equally important in the effective management of individual assets, specialized portfolios and the global investment portfolio.

The purpose of performance appraisal may be expressed as follows:

(i) To quantify historic performance and measure it against some chosen standard.
(ii) To provide explanations for good or bad performance.
(iii) To assess in quantitative terms the expected future performance, to see if the prospective performance is likely to meet the set targets.
(iv) To assist in the re-assessment of investment strategies and to point to possible adjustments.

The investor or the trustees need to satisfy themselves that the invested funds produce returns which satisfy the targets and objectives of their investment strategy and that future prospects justify the lowering or maintaining or the increasing of the level of investment in certain types of assets, portfolios or investment media. The investor and the trustees will also be interested in the effectiveness of the management of investment assets and portfolios – i.e. in the efficiency and effectiveness of their investment and portfolio managers. Performance appraisal provides the suitable basis for these requirements.

On the other hand, on the operational level, the portfolio manager needs a tool with which he can evaluate the performance of the investment asset in his care and also analyse the effectiveness of his own decision-making in retrospection. The portfolio manager's objectives in analysing investment performance are as follows:

(i) To monitor the degree of achievement of the targets set in terms of returns and risks.

(ii) To draw conclusions from the analysis of historic performance figures.

(iii) To plan future action in order to achieve maximum returns at an acceptable level of risk by making the appropriate adjustments to his portfolio through acquisitions and disposals.

The National Association of Pension Funds adopted the following definition of the objectives of the performance appraisal of property portfolios:

(i) To compare the funds performance with that of other funds.

(ii) To measure the relative performance of different types of property.

(iii) To measure the relative performance of property against competing forms of investment.

(iv) To judge the performance of property advisers.

The above definition covers the requirements of the majority of institutional investors who have substantial property investment portfolios as part of their global investment portfolio.

On the operational level, the manager of a property portfolio needs performance appraisal for the following reasons:

(i) To produce evidence for periodic reports to the investors or trustees who may use the performance statistics for inter-fund comparisons, or for the justification of further commitment of funds to the property portfolio.

(ii) To evaluate the achievements against the target set for them by the investor or the trustees and to produce performance indicators for their portfolios which could be compared with the performance of equities, index-linked gilts or some other general economic indicators such as the rate of inflation.

(iii) To provide the information base for operational investment and management decisions in order to maintain or improve the future investment performance of the property portfolio and its components.

11.3 THE EVOLUTION OF PORTFOLIO PERFORMANCE APPRAISAL

With the growth of investment funds, investors became increasingly concerned about the performance and the efficient management of their investment assets. They soon realized that even a small difference in the skill and competence of management could mean substantial variations in investment returns. Investors wanted to be reassured that their investments earned the highest possible returns whilst giving the maximum protection against risks. Important decisions to switch assets amongst funds or changes in management required reliable methods of comparison. Portfolio performance and the performance of management are different entities, both requiring appropriate standards of comparison.

The purpose of this section is to trace the mainstreams of the evolution of portfolio performance appraisal and its conceptual thresholds since the Second World War.

In the late 1940s attention was concentrated on the comparison of the behaviour of equities and gilt-edged securities, attempting to show how the average equity investor had fared in comparison with the gilt-edged investor.

The first, and most significant, historical and conceptual milestone was the publication of Markovitz's pioneering work,[1] establishing the foundations of modern portfolio theory. Although Markovitz concentrates on the problems of rational portfolio selection, his work created a completely new attitude towards portfolio performance and efficiency.

Until the advent of the Markovitz approach to portfolio selection, the performance of portfolios were measured by the rate of return in absolute terms, over a specified holding, or accounting period. No attempts were made to include risk as it was considered virtually impossible to define, let alone to quantify. It was Markovitz's two-dimensional approach which established the proper status of the risk dimension of portfolio performance.

During the early 1960s, two distinct approaches emerged to portfolio performance measurement. One approach remained closely linked with the theoretical concepts of portfolio theory, whilst the other, the pragmatic approach, was born as the result of the immediate, practical needs of the investors for a 'universal' method of performance measurement. The practical approach was favoured on both sides of the Atlantic. Whilst the Americans were involved with empirical studies on a massive scale, their European colleagues concentrated on concepts and definitions. It is interesting to note that the two earliest, complete performance measurement methods, one presented by Gilland[2] in the UK and the other created by Dietz[3] in the USA, are still the backbone of the currently used, practical methodology. Research activity in both mainstreams continued through the 1970s and up to the present.

The use of the portfolio selection models should produce 'efficient portfolios' with a better performance than the 'naïvely' selected portfolios. An immense quantity of research publications produced no conclusive evidence to prove this hypothesis. The lack of success of this approach has been explained by the immense difficulties associated with the proper definition, quantification and measurement of risk.

Sharpe[4], Traynor[5] and Jensen[6] constructed models to measure portfolio performance. All these measures were derivatives of modern portfolio theory. In the 1970s these performance measures were used in appraisal of the performance of mutual and pension funds, using statistics reflecting the funds behaviour over long time periods. The main objective of this research was to find out if managed funds produced significantly superior returns than the market. The results of these investigations were consistent: managed funds did not, on average, perform better than the market.

Malca[7] also examined the relationship between risk classes and investment performance, between performance and diversification, portfolio turnover and portfolio size. No statistically significant relationships were discovered and conclusions were that superior performance could be explained by chance alone. Active management of portfolios did not achieve better results than the 'buy and hold' investment strategies.

At first glance, the performance appraisal methodology, based on portfolio theory, has little relevance in the measurement and analysis of the performance of property portfolios. The main reasons for this are that property portfolios are assembled in a very inefficient market environment and the individual property assets have very different risk – return characteristics from those of stock market securities. Nevertheless, as the property investment market becomes more efficient, the revision of this approach for the performance appraisal of property portfolios may be justified.

The pragmatic approach to portfolio performance appraisal dates back to the publication of Gilland in 1962. The Third Congress of the European Federation of Financial Analysis[8] discussed the practical methods for portfolio performance measurement and agreed on some fundamental definitions which were recognized as the prerequisites for a unified measurement methodology. In the USA, Dietz studied the performance of pension funds and his methodology used in that early study has remained the basis of portfolio performance appraisal ever since. Dietz considers performance in the following four aspects:

1. The overall rate of return.
2. The volatility of rate of return as a proxy for risk.
3. The selection of securities of the fixed and variable return types as portions of the entire portfolio.
4. The timing of investments between the fixed and variable return type of securities.

In 1964, the Bank Administration Institute in the USA sponsored a major research programme to unify and standardize the measurement of pension fund performance for the purpose of inter-fund comparison. Their report and recommendations[9] were published in 1968.

The BAI report is significant and valuable for several reasons. First, it has endorsed Dietz's methodology which recommended the two-dimensional, risk–return approach to the measurement of investment performance. Secondly, it consolidated some important definitions of the rates of return, establishing a standard method to measure the rate of return in absolute terms. Thirdly, it recognized the immense problems associated with the definition and measurement of risk, and offered some practical solutions to these difficult problems.

Although the BAI approach has been accepted in the USA, its rationale had not been taken up in Europe for performance measurement until 1970.

The European pragmatic approach to portfolio performance appraisal was reflected in Conglong's article[10] in 1967, and in the following year in Brew's article[11]. Both articles approved portfolio performance measurement on a single-dimensional basis, considering returns only. Conglong raised the question of the acceptability of the use of an index as a yardstick, against which portfolio performance could be measured. Brew, on the other hand, discussed the method in which the performance of the portfolio is compared with the portfolio of a unit trust, although he also discussed the use of the various index methods.

At the start of the 1970s, stockbrokers and actuaries began to realize that the increasing demand for reliable investment performance measurement could be satisfied by offering commercially viable services using the fast-developing computer technology. In 1970, the Society of Investment Analysts set up a Working Group, with the specific objective of ironing out a generally acceptable approach to portfolio performance appraisal in a UK setting. The report is a milestone in the British scene as it tackled most of the problems of performance measurement for portfolios containing assets in the equity and fixed-interest sectors.

The recommended performance measurement method involves the calculation of the time-weighted rate of return for the actual fund or portfolio, and also for a notional fund. The notional fund is created by assuming that investments would be made by buying units in an appropriate index, in proportions requested by the investor. The skill of the manager could be evaluated against the profit or loss made between the actual or notional funds. The method relies on the availability of suitable indices. The Working Group recognized the immense problems associated with the risk dimension of performance and strongly recommended more research in that area. Since the publication of the report, the recommended methodology has been gaining acceptance and practical performance measurement procedures consolidated.

Interest in property investment performance has increased steadily through the 1970s, parallel with the spectacular increases in the amount of money invested in the property sector each year. Performance measurement of property investment and property portfolios has initially been offered by a number of the leading firms of surveyors and the evolution of these performance measurement systems will be further discussed in Chapter 14.

11.4 PROPERTY PORTFOLIO PERFORMANCE

At first glance, property as an investment medium is similar to other investment media. Investors strive to maximize the returns from their property assets and portfolios, whilst attempting to minimize the risks involved. The process of investment is also similar: the acquisition of property investments involve capital commitment and, in turn, the asset acquired produces an continuous income flow in the form of rents, and at the end of the holding period the market value of the asset can be converted into liquid capital. When property is viewed in this manner, there seems to be no reason why property should be treated differently than other kinds of investment as far as performance is concerned. Nevertheless, the investment characteristics of individual property assets, property portfolios and the property markets are such that their performance appraisal had to be carried out in a special way, separated from other investment assets and portfolios.

Since the beginning of the 1970s, institutional investment portfolios became the subject of regular performance appraisals. Property holdings were usually excluded from such appraisals as they caused grave complications and were considered as a major source of distortion of the overall results obtained.

The complications arise for a number of reasons, but mainly because market prices for property investments cannot be assessed on an objective and consistent basis and because the long-term nature of property investments is incompatible with the relatively short-term nature of stock market investments. Returns on property assets are also treated with great suspicion by analysts who are used to deal with returns on stock market securities.

In the early days of the 1970s the property element was a relatively small part of the mainstream institutional portfolio and the exclusion of the property element from the performance measurement appeared to be justified. As the proportion of the capital invested in property increased dramatically over the 1970s such justification was no longer valid.

At present, the separate treatment of the performance of property assets and portfolios is still the usual practice. Nevertheless, efforts are continuing to be made to incorporate the property element in the performance appraisal of the global institutional portfolio.

The elements and components of the performance of property assets are different and more numerous than those of other assets and other portfolios. For these reasons, the performance appraisal of property portfolios will be outlined and discussed in the rest of this section.

11.4.1 The purpose of performance appraisal of property portfolios

The primary purpose of the appraisal of property portfolio performance is to identify whether the targets set for the portfolio are being attained and to explain why performance was better or worse than expected.

Some traditionalists are still viewing and appraising the performance of individual properties in isolation from the rest of the portfolio, whilst others extend the appraisal to include the performance appraisal of portfolios as a complete entity.

Performance appraisal is expected to help portfolio managers to formulate investment strategies which will involve the maintenance of the most efficient asset mix and also to make correct timing decisions of acquisitions, disposals and other portfolio adjustments.

The investment performance targets may be perceived in the following terms:

(i) Target rate of returns and maximum acceptable levels of risk in absolute terms.
(ii) Targets for portfolio sectors.
(iii) Optimal asset mix or target balance.
(iv) To match or beat the performance of rival property portfolios.
(v) Performance in the context of external economic indicators.
(vi) To match or beat the performance of other competing funds.
(vii) The performance of other investment media.

11.4.2 The stages of property portfolio performance

The performance of the property portfolio is the aggregate of the performances of its components – i.e. the performance of the individual property assets and the performance of the portfolio sectors.

The property portfolio may appear to be producing a satisfactory overall return but some of the individual components may be alarmingly poor performers, whilst the other components will be exceptionally good in producing returns. It is important therefore that the strengths and weaknesses of the portfolio are identified.

The performance appraisal of the property portfolio is a three-stage process, commencing with the appraisal of the performance of individual properties and proceeding towards the appraisal of the portfolio itself through the appraisal of the various sectors of the portfolio.

Figure 11.1 illustrates the various stages of the performance appraisal of the property portfolio.

The separate stages of the appraisal enable the identification of the contribution of the component assets to the overall performance characteristics of the portfolio, so that appropriate actions may be taken to remedy shortcomings and to improve overall performance.

It is important to consider, at this point, the length of the time periods over which the performance measurement of property portfolios should be carried out. There is increasing pressure to carry out performance measurements of property portfolios over similar time periods as the performance

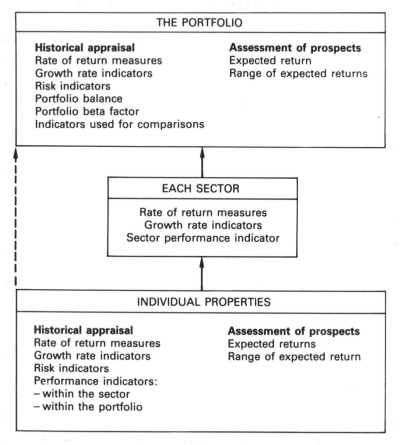

Fig. 11.1 The stages of the performance appraisal of property portfolios.

measurements of stock market portfolios. The quarterly appraisal of stock market security portfolios is becoming a standard, and property portfolio performance appraisals are expected to fall in line with this practice.

Property has always been regarded as a long-term investment medium, in which transactions and market movements are considerably slower than in other investment markets. It is therefore inappropriate to measure property portfolio performance over short time periods. The practice of using quarterly performance periods requires the quarterly revaluation of property portfolios. This can involve a horrendous computational load and also increases the subjective element in performance appraisals. This, in turn, will reduce the reliability of the results obtained. Meaningful performance appraisals should span a number of years instead of a few months. There are no hard and fast rules for the frequency of property performance appraisals, but the annual appraisal of performance appear to be reasonable practice.

11.4.3 Components of property portfolio performance

Property assets produce returns in the form of rental income and capital growth. The total return is an aggregate of these two return components, and it is important to identify and analyse which return component is dominant.

The income component of return is usually the only realized return component during the holding period of the property asset as the capital component of the total return can only be realized on disposal. Figure 11.2 attempts to illustrate how the total return on the portfolio is built up from the income and capital returns of the individual properties and portfolio sectors.

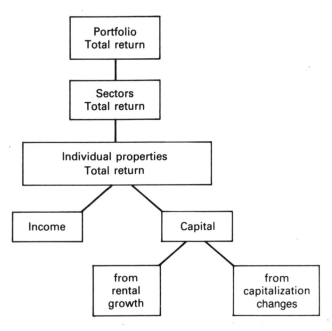

Fig. 11.2   The total return on a property portfolio and its components.

The **income return** is hard, objective evidence. The **capital return**, on the other hand, is assembled from a number of subjective assessments regarding future income, rental growth expectations and anticipated changes in yields.

The **total return** on the asset, sector and the property portfolio is computed in the usual manner, taking into account changes in capital value, net income received and additional capital in- and outflows, together with the total capital at work during the performance period.

It is customary to disaggregate the total into income return and capital return. The income return is usually calculated by disregarding capital appreciation, whilst return in capital terms assumes no rental income at all. The separate income and capital returns are then usually expressed as a percentage of the total return.

In addition, the **sources of performance** are also examined. **Rental growth** is not only important for contributing to the income component of returns, but also comes into the process of assessing market values and therefore it is an essential ingredient of capital value.

The **risk element** is also taken into consideration in property portfolio performance appraisal. Risk is usually viewed in terms of the probability that an individual asset, sector or the portfolio will not achieve its targets. Perceptions of these probabilities are derived from the historical evidence of the variability of the principal components of performance. The variability of incomes, the variability of growth of rental value and the variability of capital value change make up the historical evidence regarding risk.

11.4.4 Analysis and interpretation

Having produced the various figures and indicators from the measurement process, it is extremely important that the figures and the results are analysed and interpreted correctly.

Particular care should be taken to examine the trends in total return, disaggregating the total return into its income and capital components. Changes in capital value arise from several different sources which require separate explanations.

The **analysis of attribution** can be useful when the differences in the performance of the components of a portfolio are considered. Attribution analysis attempts to determine why the portfolio achieved a particular total return figure. The total return of the portfolio is disaggregated to some relevant attributes; each identified attribute of the portfolio return is the value-weighted average of the attributes of the components of the portfolio.

This type of analysis may be extended to include a number of components other than assets which influence the total return on the portfolio. Investment strategy, selection decisions, portfolio balance and the age of the properties, etc, all influence the results and their attributes may be examined.

11.5 PROBLEMS AND LIMITATIONS

The problems arise in the precise measurement of the various performance measures. Other difficulties are associated with comparisons.

There is a particular problem of accounting for the inflow of new monies. There is a serious loss of precision if the returns are not measured each time new monies flow into the fund or portfolio.

When comparing different funds, the dividend incomes and other internally generated money flows may occur at significantly different times; similarly, monies are withdrawn from portfolios at different dates. These money movements can cause significant distortions, rendering comparisons invalid.

Various funds and portfolios often have different constraints and limitations imposed on them. Such limitations can be statutory or can be operating constraints. For example, unit trusts usually prefer to invest in liquid assets, whilst charitable trusts have often had restrictions on the way the income is to be spent for charitable purposes. The relative evaluation of constrained funds can only be attempted if the comparison is done with similarly constrained portfolios or funds. The comparison of the performance of constrained funds with market indices is inappropriate.

There are also serious problems associated with the performance appraisal of mixed-asset portfolios containing a number of different assets from a number of investment media. Such assets exhibit significantly different investment characteristics in terms of the rate of return and risk. The presence of property assets in investment portfolios creates extreme difficulties as the performance appraisal of property assets require special treatment. The best way to resolve the problem of mixed-asset portfolios is to split them into sub-portfolios, representing the different investment media. Then evaluate the performance of each of the sub-portfolios separately and attempt to aggregate their performance contributions into the performance appraisal of the general portfolio.

There are a number of serious problems in the appraisal of property portfolio performance. The first and most difficult problem is associated with the subjectivity in the valuation of assets. The capital values assessed from various valuation procedures are a vital part of the computation of the rate of return.

The presence of **investment risk** means that none of the evaluation methods are precise, as most measures of investment risk are still awaiting universal acceptance.

The limitations of portfolio performance appraisal mainly affect the prospective mode of the appraisal. Forecasts about the future can only be considered reliable in the short run.

The historic or retrospective mode of the appraisal may suffer from the overwhelming volume of results produced. The number of indicators must be limited to the most important few in order to avoid confusion.

The sophistication of the appraisal must also be limited in order to allow the easy interpretation of the results presented.

REFERENCES

1. Markowitz, H.M. (1959) *Portfolio Selection*, Wiley, New York and London.
2. Gilland, A.B. (1962) Measuring ordinary share portfolio performance, *Investment Analyst*, 3, August, 30–5.
3. Dietz, P.O. (1966) *Measuring Investment Performance*, Columbia University Press, New York.

4. Sharpe, W.F. (1970) *Portfolio Theory and Capital Markets*, McGraw-Hill, New York and London.
5. Treynor, J.L. (1965) How to rate management of investment funds, *Harvard Business Review*, **43**(1), 63–76.
6. Jensen, M. (1969) Risk, the pricing of capital assets and the evaluation of investment portfolio, *Journal of Business*, XLII, April, 167–247.
7. Malca, E. (1973) *Bank Administered Co-mingled Pension Funds*, Lexington Books, Lexington, Mass.
8. European Federation of Financial Analysts Societies (1965) *Closing Report of the 3rd Congress*, EFFAS, London.
9. Bank Administration Institute (1968) *Measuring the Investment Performance of Pension Funds for the Purpose of Inter-fund Comparisons*, BAI, Park Ridge, Illinois, USA.
10. Conglong, G. (1970) The FT Actuaries All Share Index – a commentary, *Investment Analyst*, No.27, September, 3–18.
11. Brew, J. (1968) The measurement of portfolio performance – internal assessment, seminar paper, July, SIA, London.
12. Society of Investment Analysts (1972) *Portfolio Performance Measurement for Pension Funds*, SIA, London.
13. Eadie, D. (1973) A practical approach to the measurement and analysis of investment performance, *Investment Analyst*, **37**, 12–18.
14. Cocks, G. (1972) An objective approach to the analysis of portfolio performance, *Investment Analyst* **34**, 3–7.
15. Dietz, P.O. (1966) '*Measuring Investment Performance*', Columbia University Press, New York.
16. Marshall J.B. (1980) Pension fund performance – a new approach, *Investment Analyst*, **56**, 14–20.

FURTHER READING

Books

Hymans, C. and Mulligan, J. (1980) *The Measurement of Portfolio Performance*, Kluwer, London.
Frost, A.J. and Hager, D.P. (1986) "*A General Introduction to Institutional Investment*", Heinemann, London.
Sharpe, W.F. (1985) *Investments*, 3rd edn, Prentice-Hall, Englewood Cliffs, NJ.

The following is a selection from the published books and articles on this topic:

Bank Administration Institute (1968) *Measuring the Investment Performance of Pension Funds*, BAI, Park Ridge, Illinois, USA.
Society of Investment Analysts (1972) *Portfolio Performance Measurement for Pension Funds*, SIA, London.
Dietz, P.O., (1966) *Measuring Investment Performance*, Columbia University Press, New York.
Firth, M. (1975) *Investment Analysis*, Harper and Row, New York.
Markowitz, H.M. (1959) *Portfolio Selection*, Wiley, New York and London.
Sharpe, W.F. (1970) *Portfolio Theory and Capital Markets*, McGraw-Hill, New York and London.

Hall, P.O. (1981) Alternative approach to performance measurement, *Estates Gazette*, **2**(3), 216–29.

Hall, P.O. and Hargitay, S.E., (1984) Property portfolio – a selected approach, *Property Management*, **2**(2), 218–29.

Mason, R. (1975) Portfolio management, *Estates Gazette*, 236, November, 663–8.

Chapman, H., Wyatt, A. and Thompson, J. (1980) Measuring property performance, *Chartered Surveyor*, **112**, 444–5.

<table>
<tr><td>

12

</td><td>

The measures and indicators of performance

</td></tr>
</table>

12.1 INTRODUCTION

The purpose of this chapter is to focus attention to the specific measures and indicators of portfolio performance. In addition to the general requirements of performance appraisal, the special requirements and problems of the performance appraisal of property portfolios will also be examined.

Performance measurement is important because it enables fund and portfolio managers to assess how well their portfolio is performing and whether remedial action is necessary. However, in order to assess performance it is necessary to have a clear understanding of the principles and problems involved in the production of various performance measures and indicators. The validity of the measures and indicators must be judged against clearly defined criteria.

There is still a lack of understanding and trust of performance appraisals on the part of investors. Although theoreticians have been busy to make performance measurement systems more and more comprehensive and sophisticated, the customers (i.e. the investors) seem to lag behind in understanding and appreciation. There is a real danger that, due to its very sophistication and the production of an overwhelming number of different measures and indicators, performance measurement will become counterproductive and confusing.

Efforts must be made to educate potential users and to restrict the number of measures and indicators of performance to an absolute minimum.

12.2 WHAT IS TO BE MEASURED – GENERAL REQUIREMENTS

The performance of an investment portfolio is an aggregate of several elements. The two main elements are the return and risk.

The rate of return has always been considered as the principal element of performance, whilst risk because of the difficulty of its definition and measurement has tended to be neglected. Performance appraisal needs to concentrate on both elements of performance in order to identify the efficiency of the investment portfolio and the effectiveness of the portfolio or fund manager.

The measurement of past performance must include the following:

(i) Historic returns on the portfolio and its components.
(ii) The historic evidence of the exposure to risk.

The appraisal or performance must also extend to the identification of factors affecting performance. Attempts should be made to express the affects of these factors quantitatively and to project the future movements in these factors to aid decision-making.

Performance appraisals are carried out in absolute terms or relative to some chosen external benchmark or standard. The absolute measures of the principal components of performance are difficult to interpret in isolation.

External comparisons are much more useful, provided that there is something against which comparisons may be made; the comparisons are usually made with:

(i) The performance of the market in which the majority of the assets contained in the portfolio are traded. The performance of a market and its sectors are expressed by the appropriate indices.
(ii) The performance of benchmark or notional portfolios. The make-up of these portfolios is governed by some agreed conventions.
(iii) The movements of the economy as expressed by some chosen economic indicators.

Internal comparisons are essential for maintaining the efficiency of the portfolio; the following comparisons are made internally:

(i) Return on individual assets with the return on their sector.
(ii) Return on individual assets with the return on the portfolio.
(iii) Sector returns with portfolio return.
(iv) Risk measures of individual assets with the risk measures of sectors and the portfolio.
(v) Returns on individual assets, sectors and the portfolio with target returns.
(vi) Current portfolio proportions or balance with the optimal or target balance.

The attribution of the portfolio and its components may also be analysed to determine how performance was influenced by either the attributes of the components of the portfolio or by the various decisions about investment policy or about actual acquisitions and disposals during the portfolio period.

12.2.1 The performance period

The **performance period** is the time interval over which the performance of the portfolio is to be measured and analysed. The appropriate performance period must be defined to match the long- , medium- or short-term nature of the investment characteristics of the portfolio. The performance period must also be in accord with the requirements of the investor.

The performance period is subdivided into a number of sub-periods. The performance period usually spans a number of years, whilst the lengths of the sub-periods are months or quarters.

12.2.2 Portfolios of stock market securities

The principal measure of portfolio or fund performance is the rate of return achieved over a performance period.

Usually two types of measures of return are used: one is the **total return** on a portfolio or fund, whilst the other is the so-called **excess return**, which will be the difference between the total return and the return achievable on a benchmark portfolio.

The risk element of performance is a source of many problems. It is usually treated in terms of the various risk factors which seem to be associated with the level of performance of the portfolio. The factors to be measured are as follows:

(i) The variability of the rate of return on the portfolio and its components.
(ii) The volatility of the rate of return on the portfolio and its components. This is measured by the beta coefficient.
(iii) The level of diversification of the portfolio measured by the portfolio balance.

The **variability** of the rate of return is measured in terms of standard deviation of historic sub-period returns. The **volatility** of portfolio returns is measured by the historic beta coefficient of the fund or portfolio.

12.2.3 The property portfolio

The system of performance measurement and the indicators used must be in accord with the market which generates the risk–return relationships or trade-offs. In the case of property portfolio performance measurement, the appropriate indicators and measures of the two principal elements of investment performance must be related to, and derived from, the property investment market.

The measurement of property portfolio performance may be carried out in absolute terms or relative to the performance of other portfolios, the property market itself and investment opportunities in other investment media.

The principal measure of performance remains the rate of return. The measurement of the return presents relatively few problems. The discounted cash flow internal rate of return is now a well-understood, absolute measure. Problems arise, however, when the performances of different portfolios are to be compared. The source of the problem is that portfolio managers do not have control over the timing of flows of funds into and out of their portfolios. The **time-weighted rate of return** is considered the appropriate measure for the purposes of comparison.

The other element of portfolio performance, the risk dimension, is still posing theoretical and practical problems as there is no perfect and well-defined measure of risk available and the property market is not yet endowed with universally accepted overall market index.

The risk element of the performance of the property portfolio is measured by considering the factors listed for stock market security portfolios but, in addition, the following items are required:

(i) Historic evidence of the growth of rental values.
(ii) Time series of actual net incomes.
(iii) Historic evidence of the growth of capital values.

12.3 DATA AND INFORMATION REQUIREMENTS

Reliable data and information is required for the measurement of both elements of performance. Most of the historic data is available, although not in a suitable form. The same data will be used for the computation of the rates of return and for the various risk indicators. Information and data will also be required regarding the external yardsticks or benchmarks used.

All rate of return calculations require accurate inputs of cash flow data and inputs of the capital value of the asset at the beginning, and at the end of the holding period. For the computations of the TWRR and the T&MWRR, the capital value of the asset will be required for the end of each sub-period. This could be a heavy burden, particularly if the sub-periods are relatively short, as revaluations of the asset will have to be repeated frequently.

Further problems arise from the exact dating of the cash flows within the holding period and also within the individual sub-periods. However, if the sub-periods are relatively short, the assumption of dating the cash flows and revaluations to coincide with the date of commencement of subsequent sub-periods would not involve an appreciable loss of precision.

The data requirements of the performance analysis of property portfolios present no insurmountable difficulties. Portfolio managers keep detailed accounts of periodic incomes and outgoings, together with a record of all periodic revaluations of the properties. The following data should therefore be available:

(i) The capital value of each property at the beginning and the end of the period of analysis. These capital values should be an estimate of open market values.
(ii) The incomes and outgoings, including additional capital expenditure during the period of analysis.
(iii) Rental values and rental value changes during the period of analysis.
(iv) Rental income projections.
(v) Details of the lease structure of each property in the portfolio (e.g. length of leases, rent review dates, etc.).

Provided the above data are available, and a suitable time period is set, the historic performance of the appraisal of the portfolio can be completed without difficulty.

For the purposes of comparison, the time series of appropriate market index numbers and various economic indicators should also be available.

12.4 MEASURES OF RETURN – COMPUTATIONAL ASPECTS

As the principal element of portfolio performance, the appropriate measurement of the rate of return is of crucial importance. In performance appraisal the main measure is the **overall or total return**.

It is often necessary to disaggregate the total return into its main components: **income return** and **capital return**.

In the performance appraisal of stock market security portfolios **differential returns** are often computed. Differential returns are computed by comparing the portfolio returns with those of a benchmark portfolio.

The rate of return can be calculated in a number of different ways. The appropriate method of calculation depends on whether the rate of return is required to measure the performance of the portfolio in absolute terms as a separate entity or it may be required to make fair comparisons with the returns on other funds or portfolios. In other cases, it may be required to measure the portfolio manager's performance.

12.4.1 Definitions and computational aspects

All aspects of the definitions and calculations of the rate of return are to be found in Chapter 3. The parts of that chapter relevant to performance appraisal are reconsidered in this section.

The rate of return produced by capital invested over a time period is as follows:

$$\textbf{Rate of return} = \frac{(\text{Change in capital value}) + (\text{Net income})}{(\text{Capital employed over the time period})}$$

The above definition is a general one and valid over any time period, provided that the following conditions are fulfilled:

(i) The capital employed remains constant throughout the period.
(ii) The income is received at the end of the period in one lump sum.

In reality, assets may receive additional injections of capital and other cash flows at any time during the period over which the rate of return is to be determined. The timing and size of such money movements affect the rate of return computed if the laws of compound interest are properly taken into account.

Considering the general definition of the rate of return shown above, it is important that the timing of the movement of money into and out of the project is properly taken into account. The detrimental effects of ignoring the timing of these money flows will increase rapidly as the length of the time period increases. The determination of the rate of return requires an averaging procedure, which takes the timing of the money movements into account according to the laws of compound interest.

There are a number of ways this can be achieved. Averaging can be done on the basis of the **discounted cash flow method** or through the use of the **geometric mean**, which reasonably approximates the exponential characteristics of the compounding process.

The rate of return can then be computed over each sub-period, and the sub-period rates of return are aggregated into an overall or Performance period rate of return.

The sub-period rate of return is computed as follows:

$$r_j = \frac{(CV_t + CV_{t-1} - CE) + (GR - PO)}{CV_{t-1} + \Sigma\, CE_i(1 - p_i)}$$

where r_j is the rate of return over a sub-period j;
CV_{t-1} is the capital value of the asset at the start of the sub-period;
CV_t is the capital value of the asset at the end of the sub-period;
CE is the total additional capital input during the sub-period;
GR is the gross income derived from the asset during the sub-period;
PO are the outgoings during the sub-period;
CE_i is a part of the total additional capital input at a date between the start and end of the sub-period;
p_i is the time expressed as fraction of the whole sub-period over which the additional capital input CE_i is allowed to accumulate interest.

The above expression assumes that accumulations will occur at simple interest rates during the sub-period.

When the income earned by the asset is re-invested, then it is regarded as an additional capital input. If the additional capital inputs are relatively small and regular, then the approximation:

$$\Sigma CE_i(1 - p_i) = 1/2CE$$

may be introduced.

Thus:

$$r_j = \frac{(CV_t + CV_{t-1} - CE) + (GR - PO)}{CV_{t-1} + 1/2CE}$$

The above definition of the sub-period rate of return is the **true or absolute rate of return** for that sub-period, provided that loss of precision caused by the approximations is acceptable. This rate of return is computed on the basis of the **weighting of the additional capital** during a time period, hence it is referred to as **money-weighted rate of return** (this aspect will be discussed later in this chapter).

The definition of the rate of return over the **holding period** is more complex. The holding period rate of return is defined as the average of the sub-period rates of returns.

The method of averaging must comply with the **laws of compound interest**. This criterion is particularly important when the holding period exceeds one year.

Two methods of averaging satisfy the above criterion: one uses the process of the discounting of the cash flows over the sub-periods, whilst the other computes the **geometric mean** of the sub-period rates of return. A third method of averaging through the **arithmetic mean** of sub-period rates of return only provides a rough approximation. The numerical results produced through the above methods of averaging will be different.

The averaging through the discounted cash flow method and using the geometric mean provide different bases to compute and interpret the rate of return. The cash flows involved are regarded either on a time-weighted, money-weighted or time and money weighted basis. These different approaches to the computations provide the absolute and relative measures of the rate of return produced by an investment asset.

12.4.2 The money-weighted rate of return (MWRR)

The **money-weighted rate of return** is defined as the interest or discount rate which equates the sum of all the **realized cash flows** and the capital value of the asset at the end of a holding period to the initial capital outlay or the capital value of the asset at the beginning of the holding period. This definition is identical to the definition of the **internal rate of return** (IRR) of the **discounted cash flow** method. The MWRR is also known as the **true rate of**

return, equated yield and **redemption yield**. The rate of return produced this way will be dependent on the size and time of occurrence of realized cash flows; thus:

$$CV_{t-1} = \frac{CV_t}{(1+r_M)^n} + \sum_{i=1}^{n} \frac{C_i}{(1+r_M)^i}$$

This equation is to be solved for r_M where:

r_M is the internal rate of return or the money-weighted rate of return;

C_i is the net realized cash flow for a sub-period i;

CV_{t-1} is the capital value of the asset at the beginning of the holding period or the initial capital outlay;

CV_t is the capital value of the asset at the end of the holding period.

12.4.3 The time and money weighted rate of return (T&MWRR)

The difference between this and the money-weighted rate of return is that the computation will take into account unrealized cash flows as well. Such unrealized cash flow items are the changes in value of the asset from sub-period to sub-period.

The computational model is the same as for the money-weighted rate of return except that C_i is the net realized and unrealized cash flows for a sub-period i.

Care must be taken when attempting the interpretation of the results as they could represent MWRR or T&MWRR.

12.4.4 The time-weighted rate of return (TWRR)

The **time-weighted rate of return** is defined as the *geometric mean* of the rates of return achieved over each sub-period contained in the holding period.

For each sub-period the rate of return is defined as:

$$r_i = \frac{CV_i - CV_{i-1} + C_i}{CV_{i-1}} = \frac{CV_i + C_i}{CV_{i-1}} - 1$$

where r_i is the rate of return for the sub-period i;

C_i is the net cashflow in sub-period i;

CV_i is the value of the asset at the end of sub-period i;

CV_{i-1} is the value of the asset at the beginning of sub-period i.

Additional capital injections are to be included into the cash flows and also into the capital at work. If the length of the sub-periods are short, the usual procedure is to include the additional capital input into CV_{i-1}.

Since:

$$1 = r_i = \frac{CV_i + C_i}{CV_{i-1}}$$

The geometric mean of n sub-period rates of return is:

$$1 + R_T = [(1 + r_1)(1 + r_2) \ldots (1 + r_n)]^{1/n}$$

$$R_T = \left[\prod_{i=1}^{n} (1 + r_i) \right]^{1/n} - 1$$

The above expression is a weighted average where the weights are the lengths of the individual sub-periods. Hence the name time-weighted rate of return.

If the holding period is subdivided into sub-periods of equal length, i.e.:

$$t_1 = t_2 = \ldots = t_n$$

then: $T = n \cdot t$

therefore all sub-period returns will be weighted equally. The computed result will not be affected by the timing of the cash in- and outflows.

12.4.5 Example illustrating the computational procedures

The performance of a portfolio is considered over a period of three years. For the sake of simplicity we use sub-periods of one year. The cash flows and capital values are as follows:

Year 1	Value at the start of the year	£ 100 000
	Net income during the year	£ 6 000
	Value at the end of the year	£ 120 000
	Additional capital expenditure at the end of the year	£ 0
Year 2	Value at the start of the year	£ 120 000
	Net income during the year	£ 8 000
	Value at the end of the year	£ 130 000
	Additional capital expenditure at the end of the year	£ 26 000
Year 3	Value at the start of the year	£ 156 000
	Net income during the year	£ 12 000
	Value at the end of the year	£ 190 000
	Additional capital expenditure at the end of the year	£ 0

Calculate the MWRR, TMWRR and the TWRR.

(1) MWRR calculations
First, draw up the cash flow stream and then carry out the usual DCF internal rate of return computations:

Period	Expenditure	Net income	Net cash flow	
0	– 100 000		– 100 000	
1		6 000	6 000	
2	– 26 000	8 000	– 18 000	
3		12 000 + 190 000	202 000	MWRR = 23.62%

(2) T&MWRR calculations
Again, draw up the cashflow stream but the unrealized gains will also be included. Then work out the rate of return using the DCF method.

Period	Expenditure	Net income	Change in value	Net cash flow
0	– 100 000			– 100 000
1		6 000	20 000	26 000
2	– 26 000	8 000	10 000	– 8 000
3		12 000 + 190 000	*	202 000

* Change in value is included   TMWRR = 33.44%
in the final value.

(3) TWRR calculations
The rate of return will be calculated for each year and then the TWRR will be computed using the geometric mean:

Year 1

$$r_1 = \frac{120\,000 - 100\,000 + 6000}{100\,000} = 0.26$$

Year 2

$$r_2 = \frac{130\,000 - 120\,000 + 8000}{12\,000} = 0.15$$

Note: The additional capital expenditure has occurred at the end of Year 2 and therefore it will be effective over Year 3.

Year 3

$$r_3 = \frac{190\,000 - (130\,000 + 26\,000) + 12\,000}{(130\,000 + 26\,000)} = 0.29$$

$$(1 + R_T)^3 = (1 + 0.26)(1 + 0.15)(1 + 0.29)$$

$$R_T = (1.87)^{1/3} - 1 = 0.23$$

$$\text{MWRR} = R_T = 23\% \text{ p.a.}$$

12.4.6 Comparison of the alternative measures of the rate of return

The money-weighted rate of return and time and money weighted rate of return are the variants of the internal rate of return. The only difference between the two is that the MWRR uses only realized cash flows, whilst the T&MWRR includes all realized and unrealized items in the cash flows. Consequently, in the MWRR the effects of the fluctuations of the capital value of the asset over the holding period will not be reflected, whilst the T&MWRR will reflect, to a certain extent, the effects of asset value fluctuations.

In both of these DCF-based approaches the cash flows are assumed to be re-invested at a uniform rate and thus the computed rate of returns may be regarded as a kind of long-term interest rate prevailing throughout the performance period.

Since the performance of the portfolio depends on the decisions about the selection and timing of the right sort of action, these DCF-based methods will give an excellent account of the overall performance of an asset and will also provide insights into the effectiveness of the management of that portfolio throughout the performance period. However, these rates of return are extremely sensitive to the size and timing of cash flows. Since the managers of portfolios do not always have absolute control over the timing and size of the flows of funds into their portfolios, the direct comparison of management performance could be misleading, even unfair, when using these rates of return as a yardstick.

The time-weighted rate of return is the geometric mean of the sub-period rates of return, and as such it provides a time profile of investment performance which reflects the effects of fluctuations of capital values, as well as the periodic cash flows.

The short-term rates of sub-period returns provide a better, direct and immediate, comparison with the contemporary, alternative investment opportunities than a long-run rate of return.

On the other hand, the TWRR will not be influenced by the timing of cash flows or the timing of the changes in capital values. The TWRR assumes re-investment at short-term rates which prevail at the time of the in-or outflow of monies. The assumption of re-investment at short-term rates has the important advantage of reflecting, more realistically, the relative values of cash flows in contemporary market situations.

Rate of return on portfolios

The rate of return on a portfolio is computed as the weighted average of the returns on its component assets. The weighting is the ratio of the value of the component asset to the total value of the portfolio; thus:

$$R_p = X_1 \cdot r_1 + X_z \cdot r_2 + \cdots + X_i \cdot r_i \cdots + X_n \cdot r_n$$

where R_p is the return on the portfolio;
    X_i is the weight of the ith asset by value;
    r_i is the rate of return on the asset i.

The choice of the method of computation of the appropriate rate of return will depend on whether the rate of return would be required in absolute terms or for the purpose of measuring the effectiveness of the management of the portfolio.

Rate of return on property assets and property portfolios

The computation of the rates of return on property assets are exactly the same as for any other investment assets. The nature of data used for the computations will be somewhat different. The capital values at the beginning and at the end of the time periods over which the returns are to be calculated will (probably) be drawn from valuations. The subjectivity of these valuations is the subject of much debate and criticism.

The realized items of cash flow (i.e. the rental income and other recorded in- and outflows) are causing no problems at all.

Property performance appraisals contain all versions of the rate of return: money-weighted, time-weighted and time and money weighted rate of return. The worries about the subjectivity of the valuation data could be reduced by reducing the need for new valuations. The money-weighted rate of return requires only two valuation inputs: one at the beginning of the performance period, and one at the end of the period. The other two methods, the TWRR and the T&MWRR, require valuation inputs at the end of each sub-period.

The **total return** on the portfolio may be disaggregated into two components: **income or revenue return**, and **capital return**.

The income return is the return on the capital employed in terms of income only, and it excludes any capital gains or losses. The income return includes all realized returns.

The capital return is the return on the capital employed in terms of capital value change over the performance period. Unless the asset is disposed at the end of the performance period, the capital gain remains unrealized.

The income and capital return are expressed as percentages of the total return for easier comparison and interpretation.

12.5 THE TREATMENT OF THE RISK DIMENSION

The **risk dimension** of investment performance is treated on two different levels in portfolio performance appraisal. On the first level, performance appraisal attempts to analyse the profile of historic risk exposure in order to assess the likely future risk exposure of the portfolio. On the second level, the efficiency of the management of the portfolio risk is evaluated in retrospection.

The main risk indicators used in performance measurement are the **variance** of the rate of return on the portfolio, whilst the **volatility** of the portfolio return is measured by the portfolio beta.

There are other risk indicators in use, reflecting the special risk characteristics of specialized portfolios. Such special risk indicators will be further discussed in the context of the performance measurement of property portfolios.

The **downside risk** is often used in the prospective mode of performance appraisal. This particular risk indicator is most appropriate when a simulation approach is used to assess the future performance of the portfolio.

The efficiency of the management of the risk dimension is evaluated by examining the diversification of the portfolio. To reflect the level of diversification the **portfolio balance** is used.

The portfolio balance is the statement of the actual, or desired, composition of the portfolio in terms of proportions, or percentages of component assets of the total value of the portfolio. The main purpose of the portfolio balance is to provide a comparison between the actual and the desired state of affairs expressed in the difference between the **actual balance** and a **target balance**. The target balance is either stated by the investor or computed by the analyst.

In conventional portfolio terminology, the stated target balance is known as the **investor's** or **trustee's proportions**.

An optimal portfolio balance may be derived from one of the selection models of portfolio theory. This procedure involves the statement of the model in terms of individual assets or sectors of investment media to reflect the composition of the portfolio to be constructed.

Once the target balance has been derived or stated, conclusions may be drawn quite easily from the comparison of the actual balance of the portfolio with the target balance.

When a comparison is made between a computed portfolio balance and the actual balance, then the efficiency of the Markovitz type of portfolio diversification can be evaluated.

In general, the risk dimension of portfolio performance is seen in terms of the various risk factors which seem to be associated with the level of performance of the portfolio; these factors are as follows:

(i) The variability of the rate of return on the portfolio.
(ii) The volatility of the rate of return on the portfolio, which is measured by the beta coefficient.
(iii) The diversification of the portfolio expressed by the portfolio balance or mix.

(iv) The vulnerability of the portfolio measured by the probability that a specified minimum target rate of return may not have been achieved.

The portfolio beta is usually calculated as the weighted average of the beta coefficients of the individual securities contained in the portfolio.

Alternatively, the portfolio's historic beta coefficient is obtained by regression techniques, where the period returns are plotted against the returns on a suitable market index. From the equation of the regression line, the two principal coefficients, the **beta coefficient** and the **alpha coefficient**, can be calculated. The alpha coefficient can be used to indicate differences in the risk characteristics of the fund or portfolio and a benchmark portfolio which has similar total risk characteristics.

In certain cases, risk-adjusted performance measures are used. The use of the **reward-to-variability ratio** and the **reward-to-volatility ratio** can sometimes be justified to express performance in a single figure.

The risk dimension of the performance of the property portfolio is continuing to pose theoretical and practical problems as there is no perfect and well-defined measure of risk available and the property market is not yet endowed with a universally accepted overall market index.

The risk element of the performance of the property portfolio may be treated by considering the following factors:

(i) The variability of the rate of return on the portfolio and its component parts. The indicator for the factor is the variance or standard deviation of the rate of return.

(ii) The volatility of the rate of return on the portfolio against the property investment market. The beta coefficient of a property portfolio is derived from a regression analysis which requires the availability of a reliable and authoritative property market index.

(iii) The efficiency of the diversification of the portfolio is measured by comparing the actual balance of the portfolio with a target balance. The target balance is usually supplied by the investor or the trustees.

(iv) Capital value growth and its historical variation. This aspect can also be presented as a value gradient, indicating the rate of change of capital value at a particular point in time.

(v) The rate of growth of rental value and its historic variation. Again, this could be expressed as an income gradient, showing the rate of change of rental value at a particular point in time.

(vi) The variation of the short fall of the actual incomes received against the rental value also gives an insight into the efficiency of the income production of the portfolio which is governed by the structure of rent reviews.

The problem of the subjectivity of valuations used for the assessment of capital values continues to be the main source of suspicion for non-property people who are used to the efficient provision of market values in the stock market environment.

Regarding the efficiency of the management of risk for the property portfolio, the portfolio balance needs to be examined. Property portfolios can be diversified in a number of different ways. For example, a property portfolio may be diversified on the basis of asset classes (offices, shops, industrials, etc.) and on a geographical basis (London, South-east, South, Midlands, etc.). The portfolio balance reflects diversification on these different levels, indicating the degree of elimination of the **unsystematic component** of the total risk of the portfolio.

Portfolio balance should be regarded as an important tool for the strategic management of property portfolios. Once the target balance has been derived and stated in a suitable form, portfolio activity could be directed towards the achievement of the target balance. Periodic checks of the current balance of the existing portfolio against the target could reveal the rate of progress towards the achievement of maximum portfolio efficiency.

For example, portfolio balance of a property portfolio may be monitored on the following basis:

	Target portfolio	*Actual portfolio*
Portfolio balance		
Offices	X_{OT}	x_{Oa}
Shops	X_{ST}	x_{Sa}
Industrial	X_{IT}	x_{Ia}
Agricultural	X_{AT}	x_{Aa}
Total	1.0	1.0
Portfolio return	R_{pT}	R_{pa}
Portfolio variance	V_{pT}	V_{pa}

12.6 ASSESSMENT OF CAPITAL VALUES FOR PERFORMANCE MEASUREMENT

The determination of the capital values of the assets in stock market security portfolios does not present any difficulties as price information is available on all assets traded in the stock markets on a daily basis.

Capital value assessments will also be needed for the appraisal of the performance of property portfolios. In fact capital values of property assets are regarded as the most important parameters of investment performance. Unfortunately, the capital value of a property asset is only available if the asset has recently changed hands. Otherwise, capital values have to be assessed through a process of valuation.

The investment performance of a property portfolio and its components is assessed in terms of incomes and the realized or unrealized capital gains or losses. Capital gains or losses are made or anticipated in the market and presented in terms of the change in market values.

The recommended basis for the valuation of property assets is open market value, as defined for asset valuation purposes in Guidance Note 22 of the RICS Assets Valuation Standards Committee. The definition therein regards the capital value of a property asset as the reflection of the market at a particular point in time, in the past or in the present, but it is not expected to provide a measure of investment worth.

The determination of capital values on an open market basis implies the acceptance of the current market consensus regarding estimated rental values and capitalization rates.

The traditional approach to the assessment of open market value assumes that the current rental value will be available at the subsequent review dates and no explicit growth rate is taken into consideration. The implied or anticipated growth rates, together with other uncertainties perceived by the market, are assumed to be reflected in the capitalization rates used.

The capital value of property assets is the result of the interaction of the following market determined factors:

(i) Net income.
(ii) Estimated rental value (ERV).
(iii) Lease structure or rent review pattern.
(iv) Discount rates and yields.

On the basis of the above factors, the open market value may be assessed using one of the traditional market valuation models. There is a variety of the forms of these models incorporated into various computerized portfolio performance measurement systems.

A valuation model, if to be selected for performance measurement purposes, must satisfy the following criteria:

(i) It must produce the best approximation of market value at a given point in time.
(ii) It should be robust and flexible and able to deal with different rent review patterns, multiple tenancies, freehold and leasehold interests and automated computations.
(iii) It should be durable (i.e. it should require major revision of logic and rationale only if the market and economic conditions radically change).

In practice, the valuation models used are either based on the Term and Reversion or the Hardcore Method of valuation. In recent times, DCF-based valuation models are used with increasing frequency, particularly for assessment of the market values of leasehold properties.

The valuation procedures used and the subjectivity of the determination of the data inputs, inevitably create doubts about reliability. Very little can be done to improve the competence of the valuation models used. Reliability can only be improved if the various input elements used for the valuation are chosen with extreme care to reflect closely the market consensus. Unfortunately, market concensus itself is a nebulous entity.

FURTHER READING

Darlow, C. (1983) *Valuation and investment appraisal*, Estates Gazette, London.

Frost, A.J. and Hager, D.P. (1986) *A General Introduction to Institutional Investment*, Heinemann, London.

Stapleton, T.B. (1986) *Estate management practice*, 2nd edn, Estates Gazette, London.

Firth, M. (1975) *Investment Analysis*, Harper and Row, New York and London.

Levy, H. and Sarnat, M. (1982) *Capital Investment and Financial Decisions*, Prentice-Hall, Englewood Cliffs, NJ.

Society of Investment Analysts (1974) *The Measurement of Portfolio Performance for Pension Funds*, SIA, London.

Hall, P. and Hargitay, S. (1984) Property portfolio performance – a selected approach, *Property Management* 2, 218–29.

Hetherington, J. (1980) 'Money and time weighted rates of return', *Estates Gazette*, 256, 1164–5.

Mason, R. (1975) *Portfolio management*, Estates Gazette, London.

Messner, D. and Chapman Findlay, M. (1975) Real estate investment analysis: IRR versus FMRR, *Real Estate Appraiser*, July–August, 40(4).

Newell, M. (1985) The rate of return as a measure of performance, *Journal of Valuation*, 4, Autumn, 130–42.

Bank Administration Institute (1968) *Measuring the Investment Performance of Pension Funds for the Purpose of Inter-fund Comparisons*, BAI, Park Ridge, Illinois, USA.

Sykes, S.G. (1983) The assessment of property investment risk, *Journal of Valuation*, 1(3), 253–67.

Sharpe, W.F. (1970) *Portfolio Theory and Capital Markets*, McGraw-Hill, New York and London.

The uses of various indices in performance appraisal | 13

13.1 Introduction
13.2 A review of the indices used in investment practice
13.3 Property indices
13.4 Data requirements – databanks for property indices
13.5 Use of indices in performance measurement

13.1 INTRODUCTION

This chapter is intended to establish the basics of index numbers and to examine their application and use in the measurement of the performance of portfolios.

All investors, large and small, need to examine the performance of their investments and investment portfolios in the context of the investment market. There is a need to establish a suitable yardstick against which the performance of individual investments and portfolios may be judged and to determine whether the portfolio has achieved an above- or below-average measure of success.

Modern portfolio theory postulates that the performance of individual investments and portfolios cannot exceed the performance of an efficient market in the long run. The market can be beaten in the short run by observing price movements in the market and by the careful timing of acquisitions and disposals. It is essential therefore, to have some reliable indicator of the state of the market for two reasons: first, to enable the investor to make reasonable decisions about the timing of the purchases and disposals; and secondly, to provide a historic record of market movements against which the performance of the investment portfolio could be evaluated. Various indices are used for these purposes in the investment markets.

The stock markets are particularly well provided with sophisticated and reliable market indices, whilst other sectors of the investment spectrum are seriously lacking in this respect.

The property investment market has been notoriously ill-equipped with reliable market indices, although recent developments are encouraging. The main reason for the painfully slow evolution of reliable property market indices is the fragmented and localized nature of the property market where all individual investments are unique.

Most investment managers are concerned about their own performance, and the performance of their portfolios relative to the performance of others.

In the short run their performance will be judged against an appropriate market index and their position in a league table. Hence their preoccupation of 'beating the index' or 'tracking the index'.

As it is virtually impossible to 'beat the index' consistently in the long run, the 'tracking of the index' becomes the dominant objective.

In the stock markets the use of indices in investment work and performance measurement is widespread. Various gilt, share and bond indices provide instantaneous insights into the state of the whole market or into the level of activity in the sectors of the market.

An index can reveal the short-term changes in market prices and provide the means to identify the long-term trends of prices, yields and price–earning ratios, etc. All of these are extremely useful in making timing decisions.

On the other hand, the time series of index numbers provides the means to calculate the performance of the market, or an appropriate sector of the market. The performance of the portfolio then can be readily compared with the performance of the market, or sector of the market.

To use indices effectively the appropriate index must be used as the yardstick, and a reasonable knowledge of the nature and characteristics of the various indices is required.

Property investment assets and property portfolios present a considerable complication if they are included in the measurement of performance of a mixed-asset investment portfolio. Whilst market prices of the stock market securities are available instantaneously on an objective and consistent basis, the market price of property assets are subjectively estimated by using some appropriate valuation method. These complications are such that it is best to examine property assets separately in the context of specialized property portfolios and then regard the property portfolio as a single asset in the global investment portfolio.

Although recent attempts to reconcile property with other investments are both worthwhile and commendable, it is likely that only a small proportion of all property assets will be suitable for securitization and unitization. Nevertheless, in the property investment market there is a great and continuing need for a reliable property market index. Such an index would be similar to a stock market index only at first glance, its actual nature needs to be substantially different from any of the stock market indices in order to reflect the special characteristics of the property market.

The construction of property market indices is complicated by the nature, availability and collection of data. The data requirements are such that specialized data banks need to be set up. Without an efficient and well-designed data bank, the provision of a reliable property index is virtually impossible.

Since the property market is highly fragmented, consisting of a number of local and specialized markets, an overall property index is likely to have serious limitations. Section 13.2 is a review and critical appraisal of the indices used in the performance measurement of portfolios consisting of stock market securities. Section 13.3 is devoted to the problems of the construction of property indices. The different types of property indices will be examined, in turn, and their evolution traced through the last decade. Section 13.4 focuses on the information requirements, data acquisition and management and examines the operational details of the Investment Property Databank; and Section 13.5 attempts to illustrate the use of indices in performance measurement.

13.2 REVIEW OF INDICES USED IN INVESTMENT PRACTICE

There are a number of indices in general use in investment practice. Most of these indices are reflecting market movements and the various stock markets and security markets, whilst some like the Retail Price Index are used as general indicators of the state of the economy.

Stock market indices are usually derived by taking a collection of company shares and multiplying each share price by its company's market capitalization. These products are then used to calculate an arithmetic mean, usually with a base value of 100. The movement in a large company's share price will have a proportionately larger effect on the index than the same movement in a smaller company's share price.

13.2.1 Share indices in use in the UK

The Financial Times Industrial Ordinary Index – FT 30-Share Index

This index is based on the thirty largest UK industrial companies; it is not therefore a good indicator of the overall state of the stock market. It has been published since 1935 and its main purpose is to provide a quick guide to market movements. It is based on the use of a geometric mean and in fact it is the ratio of two geometric means. It is constructed by first multiplying together the prices of all the thirty leading shares, then taking the thirtieth root of the product; the result is then divided by the amount obtained by the same calculation at the base date. Because of its geometric basis, and the limitation of using only the thirty leading stocks, it is considered unsuitable for the purposes of portfolio performance measurement.

The FT/Actuaries All Share Index

This index has been produced since 1962 on all the main sectors of the market and is appropriately back-dated; it is published daily. The 740 largest UK companies are included in this index, which is subdivided into seven main groups and thirty-four market sectors, each with its own index. As well as the index value, the estimated earnings yield, gross dividend yield, ex-dividend adjustments and P/E ratios are shown for each sector. The indices are based on arithmetic means, and the weighting a mixture of fixed and current weightings.

This group of indices represents the most comprehensive and detailed overview of the UK stock market and is eminently suitable for portfolio performance measurement work.

The Financial Times–Stock Exchange Index – FT–SE Index

This index, affectionately known as the 'Footsie', is based on the 100 leading stocks, started on 1 January 1984, with a base value of 1000. The 100 leading stocks are in fact the 100 largest UK companies and their contribution to the index is weighted by their market capitalization. The index is a weighted arithmetic index and it is continually calculated. The base values are recalculated each quarter in order to reflect alterations in the constituent stocks and their market valuations; because of its immediacy, it can be regarded as the best guide to the short- and long-term movements of blue-chip shares.

Fixed-interest indices

The main purpose of this group of indices is to provide an insight into the yield structure at a particular point in time, together with the historic record of fixed-interest securities. The indices are widely used to provide a standard against which the performance of the portfolios containing fixed-interest securities may be evaluated.

Indices are available for gilt-edged securities and for debentures and preference shares. For these two categories, two sets of indices are provided: one set gives a historical record of gross redemption yields, whilst the other gives the prices for the fixed-interest market, together with ex-dividend adjustments, enabling the calculation of rates of return. Consequently, this set is extremely useful for performance measurement. The indices use arithmetic means and their computations are essentially the same as the FT/Actuaries All Share Index.

The Financial Times World Indices

This contains a series of arithmetic averages with weighting by multiplying the prices by the number of shares outstanding; the base date is January 1987 at a value of 100. These indices are the best guide to the movements of world stock markets and include indices for twenty-three countries.

13.2.2 Stock market indices used overseas

The Dow-Jones Industrial Index

This is a very similar index to the FT 30 Share Index, being based on the thirty largest companies on the US stock market. Similar disadvantages to those of the FT 30 Share Index makes the Dow-Jones Industrial Index unsuitable for uses in portfolio performance measurements.

The Standard and Poors 500 Index

This is an index similar to the FT–SE 100. It is weighted by the number of shares outstanding. Started in 1942, with a base value of 10, it is recalculated continuously every few seconds throughout the working day of the stock market and includes approximately 80% of the value of all stocks quoted on the New York Stock Exchange. The advantages of the Standard and Poors 500 Index are similar to those already mentioned for the FT–SE 100 Index but with a somewhat larger coverage.

The Wilshire 5000 Index

This index covers the whole spectrum of the US stock markets and therefore is comparable with the FT/Actuaries All Share Index.

13.2.3 Other indices reflecting the state of the economy

Investment returns are of course useless without reference to the change in the purchasing power of money over the period in which they have been calculated. The above-mentioned stock market indices do not allow for inflation: in order to express the performance indicators in real terms, appropriate allowance must be made for inflation.

The Retail Price Index

This index is the most important economic indicator of the UK economy, published during the third week of any given month, giving the rate of inflation at the first day of the preceding month. A new base date was set in January 1987 for this index.

The Consumer Price Index

This index is the equivalent to the Retail Price Index in the USA. Its base date is 1967 at a value of 100, and it is published in the fourth week of each month.

13.3 PROPERTY INDICES

Attempts to provide indices for the property market were made in the early 1970s. At that time, as now, the perceived purpose of property indices was to provide direct comparisons between the property medium and other investment media. This objective remains as elusive as ever, unless a fully fledged, unitized property market should emerge at some time in the future. On the other hand, the property profession now accepts the need for a reliable benchmark of investment performance. The evolution of the Investment Property Databank and its various indices can now be regarded as best instruments to achieve the above objective. The fact remains that the property market and property investments are very different from securities markets and stocks and shares.

As far as the markets are concerned, the stock markets are centralized marketplaces with a large number of daily transactions whose details are well publicized. Price evidence is immediately available and market values can be ascertained by inspection. In contrast, the property market is fragmented into local and specialized markets, with relatively few transactions whose details are not usually revealed. No objective price evidence is available and market values are subjectively estimated through some valuation procedure. Transactions in the property market are not instantaneous, as in the stock markets, and considerable time and cost is involved in completion of acquisitions and disposals.

Under these conditions, the construction of an overall, comprehensive property market index comparable with, say, the FT/Actuaries All Shares Index remains a very remote possibility. However, useful indices and indicators can be produced to suit the special conditions prevailing in the property market. So long as the purpose and objectives of these indices is well defined within the confines of the property environment, they should prove both useful and successful. The scope of property indices is limited and care must be taken that they are used only within their particular area of validity.

There are no particular technical or mathematical difficulties in the construction of property market indicators and indices. The main problem is the availability and collection of the data needed for the maintenance of the proposed indices. The total property market is difficult to assess in its entirety as the samples drawn from it will be subject to distortions and bias; however, there is some hope in as much as the property market is dominated by institutional investors. Evidence from institutional investment activities is more readily available and easier to co-ordinate; the institutional property market can be readily assessed and indices constructed. Most of the indices currently available are based on the data derived from institutional activities. Property indices therefore should be used for the analysis of institutional investment activity, and if comparisons are necessary, they should be made with institutional investment and performance in other investment media, not the performance of the markets themselves.

The problem of the subjectivity of the assessment of market values through valuations remains a highly debated issue. It is unlikely that this particular aspect of the measurement of performance of property portfolios will ever be resolved to everybody's satisfaction.

A further issue is the bias within the property market indices caused by the capital and rental valuation carried out by a relatively small range of firms. The National Association of Pension Funds recommended that the following four procedures should be adopted:

1. A database sufficiently large to enable reliable samples to be drawn from it is essential.
2. Data should be obtained from a representative range of competent valuers.
3. The range of data about each property should be sufficiently wide.
4. The independence of the sources of data should be available for verification.

13.3.1 Currently available property market indices

There is now available a wide variety of property market indices and indicators, covering both rent and capital value changes. These indices may be classified as follows:

(i) whole fund indices;
(ii) indices derived from special index portfolios;
(iii) indices based on data drawn from special locations.

Most of the indices are disaggregated into:

(i) capital or capital growth indices;
(ii) rental or rental growth indices;
(iii) total return indices.

There are, of course, many significant differences in the database, rationale and the computational methods used and different results are to be expected. Nevertheless, all the indices reveal similar trends in market movements. Undoubtedly, indices are most valuable in providing fund managers with a reasonably clear view of the state of the institutional property market. They are, of course, extremely useful in the measurement of investment performance relative to other investors in the property market and relative to the success of institutional investment activity.

Whole fund indices

These indices are constructed from the data and information drawn from all the subscribers to a particular service. The subscribing funds are combined into an aggregate fund. The data collected refers to the participating fund as a whole, rather than data pertaining to individual properties in the investment portfolios. There is now only one index in this category, the Morgan Grenfell Laurie/Corporate Intelligence Group (MGL/CIG) Property Index.

The MGL/CIG Property Index is based on data on the performance of actual funds willing to provide information. The index is comprised of three individual indices, an index of property values (Capital Index), an index of total return and index of income return. These three indices are produced on the basis of information which covers the following:

 (i) The open market value at the beginning of a period.
(ii) Purchases and sales made during the period.
(iii) Net rental income during the period.
(iv) The open market value at the end of the period.

The actual calculation of the indices is done by linking the periodic figures together; weighting is achieved on the basis of value. The details of the computations can be seen reasonably clearly in the technical appendix of the MGL/CIG report.

Indices derived from special index portfolios

These indices are derived from special index portfolios or data on individual properties aggregated into representative portfolios. The four main indices in this category are the Investment Property Databank Indices (IPD), the Jones Lang Wootton (JLW) Property Index, Richard Ellis (RE) Property Market Indicators, and the Weatherall Green and Smith (WGS) Quarterly Property Index.

The producers of these indices are either large surveying firms who manage a substantial proportion of the stock of investment properties of institutional investors, or an organization sponsored by a substantial part of the representatives of the Property Investment Market. These producers have access to a substantial amount of pertinent data so that they can create notional market portfolios by aggregating selected properties, owned by their clients, into large representative portfolios. In fact they are able to create a model of the institutional property investment market. The difference between this approach and the Whole Fund approach is that the data is drawn from the individual properties included in the notional portfolio. The notional portfolio is then used to provide the data for the construction of the index.

Although the above indices are of the same family they are significantly different in many respects. The most significant difference is the way in which the IPD indices and the JLW, RE and WGS indices are produced. The data used for the construction of the IPD indices is provided by over 100 different firms of surveyors and collated by an independent research company. The JLW, RE and WGS indices, on the other hand, are produced from the firms' data sources using their in-house valuation and management statistics.

The main features and characteristics of the above indices are reviewed below. This review would not be complete without referring to a paper by Guy

D. Morrell: 'Property performance analysis and performance indices', which contains an excellent and comprehensive comparison of the characteristics of the major property indices.

Investment Property Databank indices.

The Investment Property Databank produces an Annual Index and a Monthly Index. The latter is of great importance as the Royal Institution of Chartered Surveyors has chosen it as the principal benchmark of the performance of the commercial property industry. The IPD produces four different types of indices covering the retail, office and industry sectors together with indices for all properties included in the databank. These indices are calculated on carefully collated data on £40 billion worth of individual properties. The type of indices of market measures are as follows:

- Index of market measures excluding transactions, developments and actively managed properties.
- Index of performance of standing investments.
- Index of performance measures including transactions.
- Index of performance of total invested assets, including transactions and developments.

The uses of these different types of indices vary considerably as illustrated in the Annual Review in Appendix 2.

The IPD also produces indices by different fund types.

The indices produced by IPD differ significantly from those produced by various surveying firms in the following respects:

(i) they are based on rents receivable rather than rents passing;
(ii) they are based on the accruals principle of accounting;
(iii) transaction and property costs are deducted;
(iv) other revenue expenditures are deducted.

All these aspects are fully explained in Appendix 2 in the Section Summary of IPD performance analysis methods.

In addition to income return, capital growth, total return and ERV growth measures, the IPD also publish average equivalent yield measures and indexed yield movements based on actual properties.

The IPD also publish an index of let land performance.

The indices are well presented in tabulated and graphical forms and are easy to follow. Appendix 2 contains details and illustrations of the various products of the Investment Property Databank.

Jones Lang & Wooton Property Index

The Jones Lang & Wootton (JLW) Property Index goes back to 1967, although first published in 1978. The index is based on data drawn from the special

JLW Property Index Portfolio, which contains properties managed by JLW on behalf of their clients. The index portfolio is designed to represent a typical actively managed institutional portfolio. The properties for this portfolio are selected in a random fashion, but the typical regional spread of institutional investment is attempted to be replicated. Capital growth, Income Growth and Total Return indices are published quarterly. The method of calculation of the indices is of the chain-linked type. Further details of the index are to be found in a JLW explanatory note produced in 1984.

Richard Ellis Property Market Indicators

The Richard Ellis Property Market Indicators are based on data drawn from properties from over thirty different institutional property portfolios in which the firm was involved. The index was launched in 1978; the aggregate portfolio yielding the index was made up from some 850 properties in 1987 and covered the retail office and industrial sectors. The aggregate portfolio was not intended to reflect a typical managed institutional portfolio, but to show the changes in capital and rental values caused by supply and demand. Separate indices are produced for capital growth and rental growth for all three sectors and for all properties.

Weatherall Green and Smith Quarterly Property Index

The Weatherall Green and Smith Quarterly Property Index has been available since the end of 1979. The index is based on a portfolio which had a capital value of £475 million, in 1987, containing 279 properties covering the retail, office, industrial and miscellaneous sectors. The index contains indices of capital value and rent received, together with an index of total return. An indicator of the reversionary potential of the properties is also provided. Although this index is computed in a similar manner as the JLW Property Index, its data characteristics are different: sale prices are used instead of valuations at each quarter and include properties which were subject to development, refurbishment or lease restructuring.

Indices based on data drawn from special locations

The purpose of these indices is to reflect the changes in rental levels. They are not referred to specific properties, but 'rent points' were selected in regions of the country and samples of rents collected and weighted to reflect an institutional portfolio. There are, of course, some variations in the manner in which these indices are constructed; but the objective is to trace the market movements of rental incomes in the context of institutional investment activity. The three most important indices in this category are the Investors Chronicle/Hillier Parker Rent Index, the Healey and Baker Rental Growth Index and the Morgan Grenfell Laurie/Corporate Intelligence Group Average Rent Index.

The Investors Chronicle/Hillier Parker (ICHP) Rent Index is based on samples collected from sampling points at 1293 locations. These locations were selected to represent the type of property of greatest interest to institutional investors. The index is published every six months and the sectors are weighted in order to achieve the ideal spread of an institutional portfolio which is at the moment 48% offices, 37% shops and 15% industrials.

The Healey and Baker Rental Growth Index is similar in concept to the ICHP Index, assuming prime institutional property in the most desirable locations. The sample is drawn from locations in towns and cities which appear to be favoured by institutional investors. An index is then calculated for each location. These locational indices are then aggregated and averaged to form an index for each property sector. Equal importance is attached to rental growth in all selected locations.

The Morgan Grenfell Laurie/Corporate Intelligence Group Average Rent Index is an addition to their capital value index. It is not a true index as rental values are collected from a number of towns from local agents, in order to reflect the average rents achieved; it is published twice each year.

The above review of the available property indices reveals that there are variations in concept, style and format of the various indices, and although there are some disagreements between them, they accurately reflect the state and trends of the property market. The indices, particularly the more recent ones, are becoming increasingly sophisticated and more reliable, mainly because great care has been taken to use the greatest possible spread to represent all major sectors of the property market. The data collection and the quality of data, and the management and maintenance of data banks, are the main contributors to the increasing reliability of property indices. There are still many issues concerning the limitations of property indices and until recently, the search was on to produce a single index of property performance right across the whole spectrum of property investments. The creation of a single index derived from the various indices produced by the major surveying firms and other organizations, has proved to be a mathematical and technical impossibility. Instead, the IPD Monthly Index has been chosen as the industry standard benchmark to reflect the performance of the whole commercial property market. The RICS has chosen the IPD Monthly Index apart from its technical excellence, on the basis that it is the largest index and it is already well quoted in the Press. The indications are that as time goes by, even more firms and institutions will be prepared to subscribe their data to the IPD.

13.4 DATA REQUIREMENTS – DATABANKS FOR PROPERTY INDICES

The reliability of any index is entirely dependent on the representative nature and accuracy of the data used. An index will be only as good as the information on which it is based; therefore it is necessary to judge the data destined for use in index construction against the following criteria:

(i) Data should be relevant.
(ii) Data should provide adequate coverage.
(iii) The sample size of data should be statistically adequate.
(iv) The data should be objective.
(v) The data should be free from bias.

In the construction of stock market indices, most of the above requirements are easily satisfied. The homogeneous nature of the stock market, and the immediacy of information and large number of transactions, provide an ideal provision of data for index construction.

Unfortunately, the situation is completely different in the property invest-ment markets. Undoubtedly, the principal factor creating most of the difficul-ties for the compilers of property indices is the heterogeneous nature of the property market. Virtually all the primary sources of data – the individual properties – are unique, and for this reason, averaging and generalizations can be extremely misleading.

The main problems in property data collection arise from difficulties in providing adequate coverage in terms of type, value and geographical location, and unless access to property details is available to a large number of proper-ties, it is practically impossible to produce statistically significant samples.

An immediate and fairly good guide to the reliability of an index is the number of properties included in the survey, or the size of the portfolios used for index construction. The larger the number of properties, the greater is the coverage. On the other hand, it is important to maintain the relative size and importance of the various property sectors.

Regarding the need for objectivity, property data remains subject to criti-cism as market values are subjectively determined through various methods of valuation. The problems associated with the capital valuation of property are the direct result of the nature of the property market and the uniqueness of individual properties and the dearth of transactions relative to the total stock of property.

Bias in the collection of property data could be controlled by the different number and type of sources from which the data is collected. The recognition of the importance of data and its proper treatment has been well demonstrated in the painstaking approach adopted by the producers of the various property indices. Property data is collected either using the time series approach or the cross-sectional approach.

The **time series approach** traces the history of a fixed group of properties or portfolios. For example, rental evidence is collected periodically, including rents collected, rental values and dates of change of rental levels. A simple 'growth' index can easily be constructed from such data. The data drawn from a fixed group of properties or funds is regarded as a closer representation of the behaviour of actual assets and portfolios in the long run. Sudden major changes in large institutional portfolios are relatively rare and the time series approach of data provision can facilitate the tracing of the affects of ageing of the components of portfolios.

The **cross-sectional approach** uses selected locations instead of a fixed group of properties. The Investors Chronicle/Hillier Parker Rent Index uses this method of data collection because it is believed that by so doing data is prevented from becoming unrepresentative, and institutional investment activity is more closely represented in the short run.

The samples collected must also be presented so as to reflect the structure of the typical, desired or actual, investment portfolio. The correct weights must be used to represent the typical portfolio split in terms of geographical spread, type, value, size and age.

The data and information collected is further required to be stored in such a way as to provide easy access for perusal and for various analytical procedures. Recent developments in the construction and management of databases has made possible the setting up of large, efficient computerized property databanks. The databases set up for the production of property indices can also be used for a variety of different analytical and reporting procedures.

APRIL 1992
SUMMARY

	Retail	Office	Industrial	All Props
April 1992				
% change				
ERV growth	− 0.2	− 1.1	− 0.5	− 0.6
Capital growth	− 0.1	− 1.0	− 0.5	− 0.5
Total return	0.5	0.3	0.3	0.2
April 1991–1992				
% change				
ERV growth	− 1.5	− 12.5	− 3.1	− 6.2
Capital growth	− 1.0	− 13.4	− 0.9	− 5.9
Total return	6.3	− 6.4	8.4	1.8
End of April 1992				
Equivalent yield	8.8	9.9	11.1	9.7
Yield shift on previous month	− 0.02	0.03	0.01	0.01

- Further evidence of stabilization, but no sign of a rapid recovery in the wake of the election results.
- Little to choose between the sectors, though office rents are still falling 1% a month

Fig. 13.1 The Investment Property Databank Monthly Index. © IPD Limited.

The 'industry standard' of the property database is the Investment Property Databank (IPD). Its design facilitates the pooling of a vast amount of information and it allows further expansion. The inflow of data and information is well organized. The IPD employs well-trained data collectors whose tasks are clearly defined; data collected is subject to checks for accuracy and completeness.

The index data is collected from a statistically significant number of sources of institutional investors of varying types and sizes. The IPD claims that their suppliers of data represent 64% of the total property assets owned by UK financial institutions. The IPD now hold records of over £40 billion of investment properties.

The data on each property is collated and recorded by IPD and submitted to rigorous auditing.

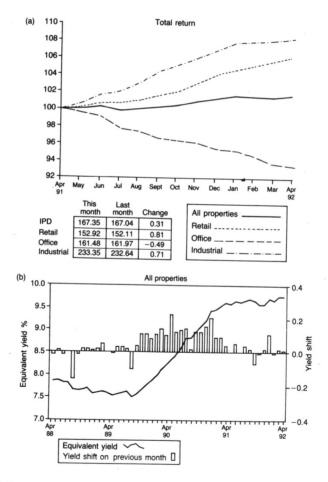

	This month	Last month	Change
IPD	167.35	167.04	0.31
Retail	152.92	152.11	0.81
Office	161.48	161.97	−0.49
Industrial	233.35	232.64	0.71

All properties ———
Retail ------------
Office — — — —
Industrial —·—·—·—·—

Fig. 13.2 The IPD Monthly Index (a) Sector Index comparisons (b) Yield Movements – equivalent yields: trends and movements. © IPD Limited.

The IPD database is used not only to compute indices and market descriptions but also for performance analyses for investor subscribers (see Appendix 2). The presentation of the IPD Monthly Index is illustrated in Figures 13.1, 13.2, 13.3 and 13.4.

	Retail			Office			Industrial			All Properties			RPI
	Rental growth	Capital growth	Total return	Rental growth	Capital growth	Total return	Rental growth	Capital growth	Total return	Rental growth	Capital growth	Total return	
87 Dec	112.67	109.20	114.73	120.20	111.30	118.60	108.47	110.36	121.92	115.24	110.16	117.05	103.7
88 Dec	136.15	128.93	142.00	150.78	143.48	162.18	133.83	144.81	173.98	142.10	136.67	153.62	110.7
89 Jan	137.60	129.81	143.53	152.23	145.30	164.95	135.75	147.46	178.24	143.64	138.16	155.97	111.4
Feb	138.69	130.47	144.81	154.46	147.68	168.40	137.20	150.17	182.58	145.27	139.77	158.50	112.2
Mar	140.61	131.14	146.13	157.28	150.28	172.12	140.91	152.73	186.77	147.90	141.49	161.15	112.7
Apr	141.87	132.41	148.13	158.74	152.33	175.25	143.86	155.36	191.10	149.56	143.26	163.89	114.7
May	143.98	133.77	150.24	160.03	154.74	178.82	145.52	158.05	195.49	151.27	145.22	166.86	115.4
Jun	145.58	134.44	151.58	163.45	157.28	182.54	148.89	161.73	201.15	153.94	147.06	169.71	115.8
Jul	146.77	135.03	152.85	165.80	159.46	185.85	152.26	164.86	206.16	156.06	148.67	172.31	115.9
Aug	147.86	135.78	154.30	167.57	161.02	188.44	154.91	168.67	212.04	157.77	150.23	174.87	116.2
Sep	149.56	135.97	155.14	170.14	162.74	191.27	158.05	171.24	216.43	160.14	151.41	177.01	117.0
Oct	151.00	135.95	155.74	171.48	163.69	193.21	159.59	172.44	219.13	161.58	151.96	178.43	117.9
Nov	152.52	135.09	155.40	173.50	163.88	194.26	161.24	172.63	220.55	163.32	151.68	178.90	118.9
Dec	154.04	134.06	154.85	176.08	164.45	195.78	162.42	172.10	221.07	165.17	151.34	179.29	119.2
90 Jan	154.39	133.18	154.48	177.53	164.46	196.65	164.73	172.30	222.57	166.40	150.99	179.70	119.9
Feb	154.83	132.17	153.96	178.18	163.30	196.13	166.20	171.89	223.31	167.17	150.04	179.38	120.6
Mar	155.92	131.12	153.40	180.15	162.82	196.43	167.68	170.27	222.52	168.69	149.12	179.13	121.8
Apr	156.77	129.94	152.69	181.05	162.18	196.54	168.89	169.92	223.37	169.65	148.31	179.00	125.5
May	157.31	127.83	150.88	181.49	157.32	191.53	169.93	167.05	220.90	170.27	144.99	175.84	126.6
Jun	158.06	126.48	149.98	182.03	155.88	190.70	170.74	165.25	219.87	170.96	143.54	174.96	127.1
Jul	158.73	125.35	149.34	182.08	153.99	189.28	172.03	163.44	218.88	171.54	142.00	173.94	127.2
Aug	159.08	123.82	148.22	182.53	151.67	187.32	172.48	161.92	218.26	171.95	140.17	172.62	128.5
Sep	159.65	122.34	147.20	183.16	149.13	185.11	173.07	160.56	217.87	172.56	138.32	171.25	129.7
Oct	160.21	120.86	146.17	183.46	147.06	183.46	173.86	158.87	217.06	173.08	136.59	170.02	130.7
Nov	160.13	119.96	145.83	183.20	145.07	181.93	174.67	157.15	216.15	173.14	135.13	169.14	130.4
Dec	160.72	118.55	144.89	183.18	142.77	180.01	175.03	155.28	215.08	173.45	133.31	167.81	130.3
91 Jan	161.24	117.66	144.56	181.73	139.77	177.18	175.73	154.05	214.84	173.25	131.52	166.51	130.6
Feb	160.99	116.80	144.28	181.50	137.83	175.70	175.69	152.66	214.40	173.06	130.16	165.74	131.3
Mar	160.96	115.62	143.63	180.97	136.17	174.60	175.95	151.52	214.33	172.90	128.81	165.02	131.8
Apr	160.94	115.13	143.82	179.85	133.77	172.52	175.80	151.03	215.18	172.44	127.60	164.45	133.5
May	160.27	114.75	144.16	177.44	132.46	171.86	175.45	150.90	216.55	171.21	126.92	164.58	133.9
Jun	160.41	114.59	144.78	176.24	130.96	170.93	175.76	151.31	218.75	170.90	126.37	164.88	134.5
Jul	160.18	114.02	144.89	174.95	128.47	168.71	176.03	150.78	219.60	170.39	125.08	164.22	134.2
Aug	160.06	113.69	145.31	173.69	127.17	168.05	175.91	151.08	221.68	169.84	124.49	164.49	134.5
Sep	159.74	113.60	146.05	172.19	125.42	166.80	175.73	151.78	224.38	169.11	123.91	164.79	135.0
Oct	159.47	113.50	146.80	170.85	124.26	166.32	175.46	151.78	226.08	168.44	123.43	165.22	135.5
Nov	159.30	113.96	148.27	169.22	123.19	165.95	174.64	151.99	228.09	167.56	123.26	166.06	136.0
Dec	159.58	114.31	149.63	166.60	121.55	164.84	174.16	152.06	229.91	166.60	122.81	166.55	136.1
92 Jan	159.47	114.24	150.44	165.32	120.48	164.49	173.53	152.31	232.01	165.94	122.43	167.13	136.0
Feb	159.17	114.12	151.19	161.93	118.96	163.53	172.16	151.33	232.27	164.23	121.62	167.15	136.7
Mar	158.83	114.12	152.11	158.94	116.99	161.97	171.19	150.40	232.64	162.76	120.72	167.04	137.1
Apr	158.57	114.03	152.92	157.19	115.81	161.48	170.39	149.68	233.35	161.83	120.11	167.35	139.2

Fig. 13.3 Index values (December 1986 = 100). © IPD Limited.

13.5 USE OF INDICES IN PERFORMANCE MEASUREMENT

In the late 1950s equities became more fashionable at the expense of the gilt-edged securities and large equity portfolios began to emerge. Prudent and risk-averse investors did not follow the fashion without asking questions about the relative investment performance of equities and gilts. The early research into portfolio performance has been concentrated largely on the historical behaviour of equities and gilts and it tried to show how the average equity investor

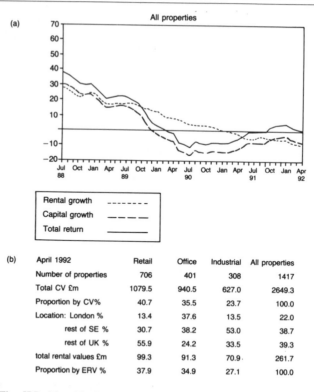

(a)

All properties

Rental growth	- - - - - - - -
Capital growth	- - - - -
Total return	————

(b)

April 1992	Retail	Office	Industrial	All properties
Number of properties	706	401	308	1417
Total CV £m	1079.5	940.5	627.0	2649.3
Proportion by CV%	40.7	35.5	23.7	100.0
Location: London %	13.4	37.6	13.5	22.0
rest of SE %	30.7	38.2	53.0	38.7
rest of UK %	55.9	24.2	33.5	39.3
total rental values £m	99.3	91.3	70.9	261.7
Proportion by ERV %	37.9	34.9	27.1	100.0

Fig. 13.4 The IPD Monthly Index (a) Quarterly trends: quarterly movements annualized (b) Index composition.

has benefited from capital gains and incomes relative to gilts. Even in those early days, the use of indices was suggested for performance comparisons.

In 1962, A. B. Gilland suggested a simple method of constructing a capital index and an income index to reflect the performance of an investment fund. The capital and income indices were maintained for a particular portfolio and compared with the performance of the stock market by using a suitable yardstick. The chosen yardstick was the Financial Times Industrial Ordinary Share Index. The internal indices and the FT index were adjusted to have the same base year, and performance could be gleaned by simple comparisons.

The rationale of using indices for performance measurement has not changed significantly during the past thirty years. The available yardsticks – i.e. the market indices – have become very much more sophisticated and specialized. The problem is now to select the appropriate yardstick from an alarming variety of indices.

The use of indices in the property medium was first suggested in the early 1970s. At that time, indices were seen as promising tools for revaluations and their use for performance measurement was suggested later; towards the end of the 1970s the indices were envisaged to be used for:

(i) Comparing the performance of property with other investment media.
(ii) Comparing property portfolios on the basis of investment performance.
(iii) Comparing the performance of the various property sectors.
(iv) Measuring the performance of individual properties relative to the whole portfolio and external yardsticks.

Internal comparisons within a portfolio do not require the availability of an authoritative property market index, but external comparisons can only be made if a suitable market index or other yardstick is available.

The comparison of the property market with other markets on the basis of a property market index must be carried out with great care to compare like with like. IPD has a formal association with the W-M Company and uses their average equity and bond portfolio returns as benchmark rather than the FTA All Share Index. (See Appendix 2, The IPD Fund Comparative Report.) The comparisons are made by using the tabulated values of the index numbers, or by depicting the relative performance on suitable graphs of indices.

The achievements of the asset or the portfolio are expressed either in terms of the percentage change observed over a period, or by converting the change observed into an index number. The change in the chosen yardstick is also observed over the same period. The relative performance of the asset or the portfolio can be evaluated and monitored by comparing the internal changes with the external ones. These processes are illustrated in Appendix 2.

FURTHER READING

Books

Edelman, D.B. (1986) *Statistics for property people*, Estates Gazette, London.
Frost A.J. and Hager, D.P. (1986) *A General Introduction to Institutional Investment*, Heinemann, London.
Hymans, C and Mulligan, J. (1980) *The Measurement of Portfolio Performance*, Kluwer, London.
Stapleton, T.B. (1986) *Estate management practice*, Estates Gazette, London.
Yeomans, K.A. (1986) *Statistics for Social Scientist* (2 vols), Penguin, Harmondsworth.

Articles and booklets

Barber, C. (1988) Performance evaluation, *Estates Gazette*, 27 August pp. 53–4.
Barber, C. (1988) Unravelling the indices mysteries, *Chartered Surveyor Weekly*, 1 September, p. 32.
Crosby, N. (1988) An analysis of property market indices with emphasis on shop rent change, *Land Development Studies*, March, London.
Eade, C. (1992) IPD index lays the groundwork for raising property's profile, *Chartered Surveyor Weekly*, **38**(3) 24–5.
Morell, G. D. (1991) Property performance analysis and performance indices: a review, *Journal of Property Research*, No. 8, 29–57.

Rowe and Pitman, (1982) Property indices: do they make sense? *Journal of Valuation*, **1**(2), 197–201.

Hager, D.P. and Lord, D.J. (1985) *The Property Market, Property Valuations and Performance Measurement*, Institute of Actuaries, London.

Healey and Baker (1986) *Property rent indices and market editorial (PRIME)*, Healey and Baker Research, London, June.

Hillier Parker (1979) *Investors Chronicle/Hillier Parker Rent Index*, Investors Chronicle/Hillier Parker, London, May.

Hillier Parker (1979) *The relationship between Shop Rents and Town Centre Size*, Research Report No. 3, Hillier Parker Research, London.

Hillier Parker (1983) *Portfolio Analysis*, Hillier Parker Research, London.

Hillier Parker (1985) *Secondary Rent Index*, Research Report No. 7, Hillier Parker Research, London.

Hillier Parker (1987) *Investors Chronicle/Hillier Parker Rent Index*, Investors Chronicle/Hillier Parker, London, November.

IPD (1992) *The IPD Annual Review 1992*, Investment Property Databank, London.

JLW (1984) *JLW Index Explanatory Notes*, Jones Lang & Wootton, London.

JLW (1987) *JLW Property Index*, Jones Lang & Wootton, London, Spring.

NAPF (1988) *NAPF Investment Committee Annual Report*, National Association of Pension Funds, London.

MGL/CIG (1987) *The MGL–CIG Property Index, 1978–1986*, Morgan Grenfell Laurie/Corporate Intelligence Group, London.

Performance measurement systems | 14

14.1 INTRODUCTION

The preceding chapters have been devoted to the examination of all aspects of the performance measurement of portfolios and the range of aspects have included the purpose, rationale and the details of the various measures used in performance appraisal. Amongst the various problems encountered, the collection and preparation of data required for various components of the analysis have presented the greatest problem. Apart from the data problem, we have also seen the complexity of the technical problems of evaluating the data and producing the conclusions in an appropriate form.

Portfolios containing property assets demand specific treatment when evaluating their performance. The problems associated with the collection and preparation of the data, required for the performance appraisal of property portfolios, are considerably more complicated than those associated with the portfolios of stock market securities. The main thrust of this chapter will therefore be directed towards the examination of performance appraisal systems specifically designed for the analysis of property portfolios.

The analysis of any portfolio, either as a whole or as a composite of its parts, requires an enormous amount of computation. If all such computations are to be carried out manually, then by the time meaningful results would be available, the critical date for the making of decisions would be well past.

Taking just one aspect of the appraisal of property portfolio performance, the periodic revaluation of a medium-sized portfolio would take several days to complete using traditional valuation methods. Little wonder that an exercise such as the frequent periodic revaluation of the property portfolio is feared by many portfolio managers.

The tedious manual procedures hampered the acceptance, or even the consideration of, serious in-house performance appraisals. The usual practice is the employment of the services of a specialist firm which through its special facilities is capable of coping with such heavy exercises. The problem with this kind of practice is that portfolio managers will receive the results of the performance appraisal after considerable delays and in a format which is not, necessarily, immediately beneficial for their operational decision-making and management activities.

The advent of the relatively cheap and abundant computer power has opened up a new way to mechanize and automate the tedious and repetitive parts of the computational procedures. It is now quite a realistic expectation to complete the computations involved in the quarterly revaluation of a medium-sized portfolio, in one day or less, depending on the form and sophistication of the required valuation report.

The investment fraternity in general, and property people in particular, were quite slow in realizing the power and potential of computers. They did not trust, and some still don't, the 'new-fangled machinery'. In addition, the early *ad hoc* approaches to computerization revealed the inherent inflexibility and limitations of the early systems. Another reason for the relatively late start of the use of computers in the property world is that people did not know exactly what they wanted or could expect from the measurement and analysis of performance.

The comprehensive appraisal of portfolio performance must be regarded as a system of separate but interdependent computational procedures. If such a system is to be computerized, all its components must be circumscribed and defined before the pieces making up the jigsaw can fit together reasonably.

Portfolio performance needs to be monitored not only to satisfy the requirements of the investor, but also to provide the base for operational decision-making and portfolio management. Performance appraisals provided by outside firms not only satisfy the need for the production of independent and unbiased reports to the trustees, but can also provide a check on the results and interpretation of in-house performance appraisals.

14.2 THE CONCEPT AND STRUCTURE OF PORTFOLIO APPRAISAL SYSTEMS

An investment portfolio is a complex system of its component parts, the individual investment assets. The portfolio, a single entity, is created to maintain and to increase the efficiency of the production of returns and the management

estment risks. A portfolio requires careful and expert management if it
achieve all the objectives prescribed for it by the investor. Some portfo-
particularly those containing property assets, require more management
ttention than others.

road terms, the management of the portfolio consists of the following:

monitoring and appraising the performance of the portfolio and its com-
onents;
ormulating plans for the adjustment of the portfolio;
Making decisions and executing plans for adjustments.

omprehensive appraisal and analytical system is an essential tool in the
ive management of the portfolio. Such a system will incorporate:

he provision of data and information regarding the behaviour of the
ortfolio and its economic environment in the past;
he facilities to measure and to analyse the historic performance of the
ortfolio and its components;
he facility to assess the capital values of the individual assets;
acility to examine the current structure and composition of the portfolio;
n appraisal framework to aid the making of selection decisions regarding
ew acquisitions and disposals;
a suitable reporting facility.

ust be clearly understood that **performance measurement and analysis**
e components of a comprehensive portfolio appraisal system. Perfor-
e appraisal is often seen as the dominant component in portfo' apprai-
owever, the other components are also vitally important.
concept of a portfolio analysis and appraisal system is outl' in Fig. 14.1.
erformance appraisal system is an integral part of the o' all system.
system divides up into three distinct time sectios, one dealing with
ast, the second dealing with the present in which the past results are
ted and policy for the future formulated, whilst the third is concerned
the future.
whole system rests on a comprehensive data and information base
requires careful maintenance and updating. Without such a data base
ining the time series of historical evidence, meaningful measurement and
sis of portfolio performance is not possible.
number of sub-systems can be identified within the portfolio analysis
m:

Data and information management sub-system.
A valuation and appraisal sub-system.
A performance appraisal sub-system.
Report generator.

he above sub-systems are interactive and interdependent.

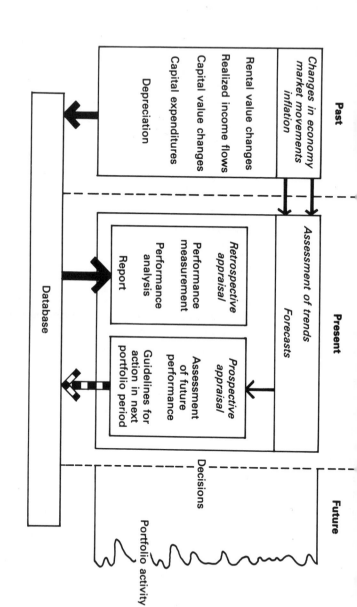

Fig. 14.1 The concept of a comprehensive portfolio appraisal system.

The following labels appear within the figure:

Past   **Present**   **Future**

Changes in economy
market movements
inflation

Rental value changes
Realized income flows
Capital value changes
Capital expenditures

Depreciation

Assessment of trends

Forecasts

Retrospective appraisal
Performance measurement
Performance analysis
Report

Prospective appraisal
Assessment of future performance
Guidelines for action in next portfolio period

Database

Decisions

Portfolio activity

Data, extracted from cash flow records and generated through the valuation of assets, is stored in the database. Performance appraisal procedures will draw data from the database, process the data and, eventually, feed back the results into the database.

The reports are produced by extracting the historic records and the results of the performance appraisal procedures from the database, together with the analyst's commentary and explanations.

14.3 SPECIFICATIONS OF PORTFOLIO APPRAISAL SYSTEMS

The selection of a suitable 'ready-made' portfolio appraisal system package for in-house use needs clearly defined criteria. Specifications for the whole system and its components must be outlined. Specifications need also to be formulated when selecting one of the portfolio appraisal services now available.

14.3.1 In-house portfolio appraisal systems

The specifications of in-house systems must be considered in terms of the operational requirements of the portfolio manager. The required sub-systems must be identified and specifications for each sub-system outlined. On the system level, the most important aspect is compatibility of the sub-systems and the ease of the data and information transfer between each module. The following sub-systems or modules are required:

1. Data and information base.
2. Performance appraisal module.
3. Valuation and appraisal module.
4. Report generator module.
5. Communication module.
6. Utilities and system maintenance module.

The **database** is usually a 'stand alone' filing system. The database is a collection of files on individual properties and the portfolio sectors. It must be structured to allow the easy access of data needed for the computational and reporting modules. It must also facilitate automatic updating, either by the computational modules or by communicating with the global database of the firm, fund or institution or by accessing large external databases.

The database requires an efficient database management procedure which will facilitate the perusal and updating of data files.

The **performance appraisal module** is to provide the automatic valuation of the portfolio assets prior to computations of the various performance measures and indicators. The assessment of the current market values of all assets are necessary to compute th ~eturns from the change of values during the last performance sub-period.

A special problem of the specification of the automatic revaluation procedure is the choice of the appropriate valuation model. The assessment of current market values is required and therefore the valuation model used must be a close replica of the procedures used in the market-place.

The valuation of the assets in stock market portfolios presents virtually no problems as market prices are readily available.

The assessment of the current market values of the assets in property portfolios presents the greatest difficulties.

The most frequently used models are those of the Term and Reversion and the Hardcore methods of valuation. There are some serious limitations of the computerized versions of these models as multiple tenancies, differing rent-review patterns, voids and leasehold assets can cause immense problems. The valuation of development properties requires different valuation models.

Although the DCF-based valuation procedures are becoming more popular in valuation practice, they are not yet a standard market method. DCF models can be used for the assessment of leasehold assets as the Term and Reversion and the Hardcore based models can only cope successfully with freeholds and long leaseholds.

The data required for the valuations are extracted from the database. It is essential that the database is updated prior to the computer runs regarding capitalization rates. The assessed current values are then fed back into the database ready for use in the computations of the performance measures and indicators.

It is extremely important that the time series of capital values, assessed using automatic re-valuation procedures, are regularly checked by external independent valuations.

The other component of the performance appraisal module is the complex of procedures to compute the performance measures and indicators. This component is expected to produce various rate of return measures achieved over the performance sub-periods and the whole performance period. There should be no problems of producing rates of return on the time-weighted, money-weighted or time and money weighted bases. The results must be checked periodically using manual or alternative computational facilities as 'bugs' in the system can cause seriously misleading statistics. Income returns and capital returns can be produced separately if required. All data required for these computations are extracted from the database. The computations will require a time series of capital values, incomes received, estimated rental values and other cash flow items.

The **risk dimension** of portfolio performance is usually explored and explained through the examination of the **variability** of the above-mentioned rate of return measures over the sub-periods and the whole performance period. The **volatility** of asset returns and portfolio returns are measured with the appropriate **beta coefficient**. The computations of the beta coefficients require the time series of the appropriate market indices. These can also be stored in the database.

The portfolio manager should monitor only the most pertinent performance measures for his/her management work in order to avoid confusion. For the instantaneous check of the behaviour of assets and portfolio sectors **internal performance indicators** can be devised. These indicators can show (very effectively) how well or how badly a particular asset is performing relative to its sector and relative to the portfolio.

Portfolio appraisal systems are also expected to give assistance to the portfolio manager in decision-making concerning the acquisition of new assets and disposals of certain portfolio assets. The availability of valuation and appraisal procedures, independent of the portfolio database, is extremely useful in the making of selection decisions. These procedures are included in the **valuation and appraisal module**.

The experimentation with the existing portfolio and the inclusion of new assets can reveal the impact of the adjustment of the composition of the portfolio. The net expected contribution of a new asset to the total return and risk characteristics can only be gleaned by such an experimentation. However, such experimentation must be carried out without interfering with the carefully maintained database.

The specifications of the valuation and appraisal module must cover the requirements for a variety of different appraisal models, which can provide the means for extensive sensitivity analyses and simulations.

The **report generator module** needs to be able to extract the information and results produced by the computational modules and produce concise reports which then can be transferred to the wordprocessing systems adopted by the firm or institution.

The Portfolio Appraisal System must also have adequate facilities to import and export data and results. A **communication module** is essential for this purpose. Communications is a rapidly developing technology and specifications must allow for flexibility and compatibility as it evolves in the future.

Portfolio appraisal systems must provide for some degree of in-house maintenance. These facilities are to be provided by the **utilities module**. The database needs to be backed up at regular intervals, to provide an insurance against the loss of data caused by system failure. Repairs may also have to be done to corrupted parts of the database.

14.3.2 External performance measurement services

External performance measurement services must be selected to satisfy the requirements of the investor or trustees and to provide independent checks for the performance appraisals carried out by the portfolio managers using their in-house systems.

The investor and the trustees want to know:

(i) How the portfolio performed in comparison with the portfolios of competitors.

(ii) How specialized portfolios, like property portfolios, performed relative to the global investment portfolio.

(iii) How exceptional performance, good or bad, is explained.

(iv) How the sectors of various portfolios performed relative to each other.

The investor or trustees do not want a deluge of performance statistics, they require clear and concise answers to the above questions.

14.4 SURVEY OF SOME OF LEADING PERFORMANCE MEASUREMENT SYSTEMS

Since the late 1960s, a whole industry of investment advisory services has grown up. Amongst these advisory services, investment performance measurement services compete with each other for the measurement and appraisal of the various investment funds, particularly pension funds. The services offered differ considerably in techniques, coverage and presentation.

The services concentrate mainly on the global investment portfolios of the funds and generally regard investment performance in terms of the success of the allocation of monies between various investment sectors and the choice of the individual investments within a particular sector. The general terminology refers to sector selection and stock selection. It is generally accepted by the investor clients of these services that the performance measurement should produce a report to the trustees on both the effects of sector selection and stock selection.

The effect of timing of movements between sectors, including cash, should be regarded as part of the sector selection. The following sectors are expected to be covered:

(i) fixed interest sector;
(ii) the UK equities sector;
(iii) overseas equities;
(iv) cash;
(v) property;
(vi) other investments.

Most of the performance measurement services analyse performance in terms of sector selection and stock selection and for this purpose some of the services use notional or model funds against which they compare each individual fund. They use various indices for the determination of the rate of return for each sector of the model funds. The indices used in this context have been discussed in Chapter 13.

The inclusion of the property sector has proved to be the most troublesome of all the sectors, and great efforts have been made – with some success – to fit the property sector into the evaluation of the investment performance of pension funds. The problem, however, is not yet fully resolved.

In order to resolve the problem of property in the performance measurement of global investment portfolios, research was initiated within the property sector by some of the leading firms of chartered surveyors during the 1970s. In 1979 and 1980, two property performance measurement systems were launched, one by Jones Lang & Wootton (JLW) and the other by Richard Ellis and Wood McKenzie Co. Both systems compute historic returns in similar ways.

The JLW system produces time-weighted rates of return to express the performance of property funds and the overall fund performance can be broken down into the performance figures of individual properties. The overall performance can be compared with the JLW Property index.

The Richard Ellis/Wood McKenzie Property Performance Service consists of an annual report supplied to the subscribing funds providing details of the portfolio analysis of the individual fund including the total return of the portfolio and the return of its component parts. The service also provides comparative protfolio analysis of the individual portfolio with other portfolios using league tables. There is also a comparison of the performance of the portfolio against various indices including the FT All Share Index.

The Hillier Parker Portfolio Analysis System provides an insight into the performance of the portfolio through the following schedules:

1. The structure of the portfolio.
2. Geographical spread by capital value.
3. Anticipated income over the next 10 years.
4. Historic rental income growth.
5. Historic rental value growth.
6. Historic capital value growth.
7. Rate of return.
8. Long-term performance.
9. Comparison of rental growth with Investor's Chronicle/Hillier Parker Rent Index.
10. Comparison with other investments and economic indicators.

In the early 1980s a number of sophisticated property portfolio performance measurement systems emerged. St Quintin's offered a comprehensive system named COMPAS, and the latest and possibly most sophisticated performance measurement service is provided by the Investment Property Databank (IPD).

The IPD offers a comprehensive performance report to the subscribing funds under the following main headings:

1. The structure of the fund portfolio.
2. The performance of the fund portfolio in terms of:
   – total return;
   – income growth;
   – capital growth.

Under these headings there is a detailed analysis of the components of performance and comparisons with the IDP portfolio are made using tables and graphs. The performance report is presented in a clear and concise form and the explanation of the technical detail of the computations are also well presented and easy to follow.

All these performance services contain a considerable amount of information on the investment performance of funds and their portfolios; however, their effective use in the management of investments and investment portfolios depends to a large extent on the ability of the potential users to interpret the results correctly.

FURTHER READING

Jones Lang & Wootton (1982) *Property Investment Performance over 20 years*, Occasional Paper, Jones Lang & Wootton, Summer.

Hillier Parker May & Rowden (1983) *Portfolio Analysis*, Hillier Parker May & Rowden Research Department, August.

St Quintin (1982) *Compas Computerised Property Appraisal System*, St Quintin, London.

Richard Ellis/Wood McKenzie & Co. (1980) *Introduction to the Property Performance Service*, Richard Ellis/Wood McKenzie, London.

Jones Lang & Wootton (1980) *Property Performance Analysis System*, Jones Lang & Wootton, January.

Investment Property Databank (1986) IDP portfolio analysis – demonstration fund, *IDP Annual Review*, June.

Hargitay, S.E. (1985) *Property Portfolio Analysis Package*, private publication, Bristol.

Frost, A.J. and Hager, D.P. (1986) *A General Introduction to Institutional Investment*, Heinemann, London.

Stapleton, T.B. (1986) *Estate management practice*, 2nd edn, Estates Gazette, London.

Society of Investment Analysts (1974) *The Measurement of Portfolio Performance for Pension Funds*, SIA, London.

Mason, R. (1980) Performance measurement, *Estates Gazette*, 256, 1091–5.

Information needs – Provision, Management

PART FOUR

Information needs and the sources of data | 15

15.1 Introduction
15.2 Property information needs
15.3 Property information sources

15.1 INTRODUCTION

Property investment analysis is basically an activity involving the use of information (input) to produce a service (output). It is therefore fundamentally important that the analyst knows what information is required and where to obtain it.

This chapter focuses on property information by identifying as comprehensively as possible all the information needs and sources required for property investment analysis. The main objective is to provide an empirical presentation of property information in all forms and at all levels. Practitioners may argue the necessity of such work but there is indeed a gap in understanding which needs to be filled.

This chapter begins with a discussion on the range of information needs for property investment analysis. The various information items are broadly categorized into: information on the economy, the property market and individual property. Section 15.3 presents the sources where such information may be obtained. Most of these secondary sources are discussed according to their type, for example, handbooks, statistical publications, newspapers, and so on.

15.2 PROPERTY INFORMATION NEEDS

It has been shown that information needs for property investment analysis can be defined in many ways. Fraser (1984) suggests that the information needed hinged on the type of interests of the property. From the investment point of view, whether a property is a reversionary freehold or a long-term leasehold can be likened to an investment with growth potential or with a fixed income. This, in principle, allows the analysis to be compared with equities or gilts, in respect to the capital market, and taking into account the state of the economy.

Jaffe and Sirmans (1981) offer a series of basic steps for the investment analysis process; the five basic steps are:

1. Identify the objectives, goals and constraints of the investor.
2. Analyse the investment environment and market conditions.
3. Develop the financial analysis and forecast cash flows from the project as well as the costs of investment.
4. Apply the decision-making criteria which will convert the expected benefits (cash flows) into a valued estimate for the investor.
5. Make the investment decision.

Although this process does not define the information needed, it identifies the action involved at each stage, which can help to ascertain when and what information is needed.

Hager and Lord (1985) define the information needed as property variable characteristics and factors affecting valuation. They group these into two main headings:internal, comprising legal, economic and physical factors; and external, comprising location, environment services and economic factors.

One point is clear from the various perceptions of information needs. The scope of information required is considerably greater than for a specific valuation. It ranges from broad macroeconomic indicators to the microeconomic aspects of the property market and then the individual characteristics of the property concerned. Property, as a unique category of investment because of its individualistic features, therefore requires even more information than other investment media. Broadly, the information needed can be categorized as follows: information on the economy, information on the market and information on the individual portfolio and property (Fig. 15.1).

15.2.1 Information on the economy

Macroeconomic information usually comes in the form of statistics published by official bodies. The importance of such quantitative data does not need further emphasis since property investment must be seen in the context of the economy. The economic environment determines such fundamental factors as interest rates, economic growth, level of economic activity, consumption and investment. Historically, economic activity tends to be cyclical. The troughs and peaks of these economic cycles have important implications for the timing of the investments. The identification of this cyclical pattern by means of certain econometric models is therefore regarded as a key information for investment analysis.

The most commonly used measure of economic activity is the Gross National Product (GNP) or Gross Domestic Product (GDP). Other statistics include

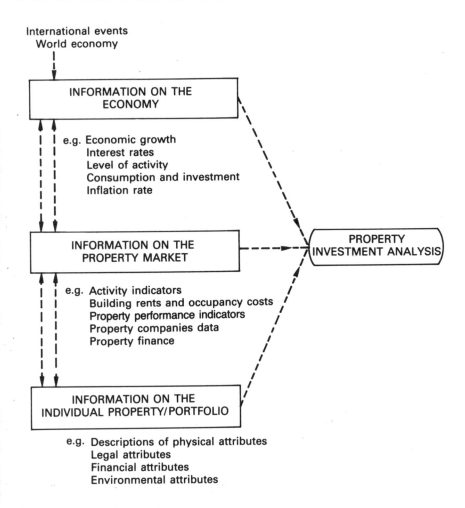

International events
World economy

INFORMATION ON THE
ECONOMY

e.g. Economic growth
Interest rates
Level of activity
Consumption and investment
Inflation rate

INFORMATION ON THE
PROPERTY MARKET

e.g. Activity indicators
Building rents and occupancy costs
Property performance indicators
Property companies data
Property finance

INFORMATION ON THE
INDIVIDUAL PROPERTY/PORTFOLIO

e.g. Descriptions of physical attributes
Legal attributes
Financial attributes
Environmental attributes

PROPERTY
INVESTMENT ANALYSIS

Fig. 15.1 Categories of information.

figures on Personal Saving, Consumers' Expenditure, and Fixed and Stock Investment. Corporate profits give a general picture of the economic prospects and, together with the GNP, are indicators of economic growth. This information is essential for understanding the state of the economy.

Inflation is a further measure of the state of the economy and has been particularly important since the early 1970s. The Retail Price Index is the basic indicator of inflation. The other indices of relevance are the Earnings Index and the Index of Producer Prices. Unemployment statistics as a factor affecting economic output have also been used more widely since the 1970s.

Information concerning money and finance has an important bearing on the level of interest rates and taxation. Money supply figures such as the M0, all cash in circulation, and M3, all cash plus bank deposits, directly influence the level of interest rates. The Public Sector Borrowing Requirement (PSBR) statistic is another useful indicator of the level of finance in the economy. A high PSBR normally means an increase in the competition for funds as a result of the financial squeeze on the private sector. The annual Budget which carries the PSBR figure will aslo spell out the changes to tax rates, one of the key factors of investment analysis. The balance of payments and the level of foreign reserves are other two important indicators of the strengths of a country's economy. They also affect the foreign exchange rate of Sterling which may be important to overseas investments. Although this information may not seem relevant to an investment analysis of a property, it is vital to an understanding of the current and future states of the overall economic environment of which the property market is part.

Various non-statistical or qualitative information about different industries, such as policy measures, competition and consumer protection, is also necessary where the investment is affected by these particular industries. The manufacturing and trading sectors of the economy are usually broken down into specific categories which then provide useful information on the state of the respective industries. The Confederation of British Industry's surveys on the various aspects relating to confidence in the economy are also relevant for gauging future prospects. Deregulation of the capital and labour markets should also be noted because of its impact on certain industries.

This information can provide a valuable insight into the present state of the economy, as well as being an indication of future prospects, based on past trends. There are many more macroeconomic statistics which are produced officially, and there are also forecasts and predictions of the important indicators made by economists and other experts. This information can be invaluable as supporting data and should be incorporated in the information armoury of the surveying firm. It should also be made available to surveyors for investment analysis, given the increasing significance of property in the overall investment market.

The value of the information of the economy, however, has to be set off against some shortcomings. One of these is that the statistics produced are almost totally national or regional aggregates. Statistics at the local level, which may be of greater significance, are often not available. As a result, local variations and spatial differences may distort the application of the information in analysing the investment potential of any local market. There are statistics such as those held by the Inland Revenue which would be very helpful; unfortunately, this and other information is not made available for reasons best known to the authorities themselves.

15.2.2 Information on the property market

Decisions concerning the timing of the purchase and disposal of investment assets require pertinent information about the state and mood of the market.

Like the stock market and capital markets, the property market is characterized by fluctuating values, depending on the investors' perceptions of prospects of returns. Some of these perceptions are based on sound economic analysis of past data but some are psychological and even irrational! In the stock market, for instance, the graphic descriptions of the 'bull' and 'bear' market are a result of the psyche of investors who often may be led by 'herd instinct'. The stock market is also described as 'efficient' because information is reflected instantaneously in the price of shares. The same, however, cannot be said of the property market since there is no central market for transactions and for generating information. Property investment analysis therefore has to rely on the gathering of up-to-date market information from all available sources. Market information concerning the immediate past is best obtained from the appropriate market indices. In the property market, apart from those generated internally within firms, the types of information available are:

1. Activity indicators – property transaction and investment.
2. Building rents and occupancy costs.
3. Property investment performance indicators.
4. Information on property companies.
5. Property finance – bank lending to property.

(1) Activity indicators.

The level of activity in the property market is the most basic indicator of its state. The information that reflects this can be found in transactions of land and existing buildings, investment in new constructions and improvements and institutions' investment in property.

Transaction data comes in three forms. First, the surveys carried out by the Inland Revenue on conveyancing in England and Wales. The surveys, limited to activities in one week of each year, are reported in *Economic Trends*, and since 1974 under the title 'Trends in Sales of Land and Buildings'. They provide information on the number and value of transactions by price range, type of property (land, residential and commercial) and by tenure. Further breakdown by region and sector of buyer and sector of seller is also available. The second form of information on the level of transactions can be distilled from censuses of production and distribution. The third is given by the treatment of land and existing building in the UK National Accounts. In the aggregate, expenditure on the purchase of land and existing buildings would net out with sales were it not for the inclusion of transfer costs in the costs of acquisition. These costs are included as part of gross domestic fixed-capital formation and appear as a separate item in these accounts as 'purchases less sales of land and existing buildings'.

Statistics of expenditure on new construction and improvement constitute the series of gross domestic fixed-capital formation compiled for the UK National Accounts by the Central Statistical Office.

Finance Statistics provides official quarterly figures on the net acquisition of the UK land, property and ground rents by institutional investors. Further analyses by institution type, such as pension funds, insurance companies, unit trust, and so on, are also available.

All these activity indicators serve to inform on the state of the property market. Whether the market is in stable or dynamic phase is fundamental to the timing of investments and is crucial to their success. It is therefore important that this information is monitored by surveying firms for use in investment analyses and other work.

(2) Building rents and occupancy costs

Levels and trends of rental values provide the basic information for the determination of income and possible growth. Information on rent can be extracted from the two main sources. The first is again the UK National Accounts, since rent is a major form of factor income and as such constitutes an important component. The rent given denotes gross receipts from the ownership of land and buildings, less actual expenditure on repairs, maintenance and insurance.

The second is the widely available market indices and reports published by a host of organizations. Most of these sources are provided by the major London-based surveying firms as part of their services to clients. The information of rent comes in the form of either rental values or as indices. Of these sources, only one is official. The data provided by Inland Revenue District Valuers give typical rental in new leases. According to Fleming (1985), 'This must be regarded as the most comprehensive source, not only because of the direct involvement of District Valuers in the property market throughout the country but also because of the information about transactions available to them from the "Particulars of Deposits" procedure for stamp duty purposes'. On the other hand, returns made by the District Valuers can be subjective. Comparisons over time are affected by the different perceptions of 'typical' and the comparisons over space are not possible because the values are for specific towns. The information is therefore indicative rather than definitive.

The private sources include: RICS – Institute of Actuaries City Office Rent Indices; RICS – FT Property Indicators; Jones Lang Wootton; MGL/CIG; Healey and Baker; Debenham Tewson and Chinnocks; Kenneth Ryden & Partners; Investors Chronicle Hillier Parker; and Richard Ellis. All these sources are, of course, different in many ways. Some of the differences are with respect to property categories, geographical coverage, frequency, period covered and nature of primary data source. Users of this information must therefore be aware of the differences and any comparisons of them must take them into account. The problems of constructing a standard index, particularly for the property market, are examined later.

Another aspect of property market information is the occupancy costs of buildings. Expenditure of the upkeep and associated operational and running costs of buildings, such as cleaning, lighting and heating, is relevant to rental income. Perhaps the best source of reference for this information is the Building Maintenance Costs Information Service, which provides subscribers with annual reports on Average Occupancy Costs Study, Costs Analyses and Study of Energy in Buildings. The information is given in terms of average costs and is available for different building types.

(3) Property investment performance indicators

In addition to rental value information, there are also a number of market indices and reports on investment yields, returns and capital values. As argued earlier, this kind of information is increasingly taking on a more prominent role in performance measurement. Current levels of capital values and returns and past trends are both fundamental information input as well as output. However, there are also similar differences between sources which must be taken into account: Inland Revenue District Valuers, JLW, MGL/CIG, Healey and Baker, ICHP, Richard Ellis and RICS-FT Property Indicators. The information comes in various forms of yields and capital values. It is also normally subdivided into different property categories and geographical coverage. Each source derives its data from different primary sources and also differs in the frequency of production and the period it covers.

(4) Information on property companies

Information on property companies is not only important for investment in the shares of the companies, but also because their activities have a great impact on the property market. The earning yield, the dividend yield and the price–earnings ratio are some of the statistics which are available in newspapers, particularly the *Financial Times*; there is also information about the activities, future plans and objectives, profits, etc. in the annual reports of the companies.

(5) Property finance – bank lending to property

The earlier section on institutional investments in property did not consider one of the major participants of the property market, the property companies. This is because there are no official statistics on property companies' acquisition and involvement. However, as bank lending to property companies is a substantial source of funds (apart from public companies which have alternative means, especially equity funds), a reasonable indication of trends in acquisitions can be gained from trends in bank lending.

Property has traditionally been regarded as a secure form of collateral for banks and other lending institutions. Information on banks' lending to property has therefore profound implications for the current state of the property market, as well as for future prospects. The correlation between the significant rise in bank lending in the early 1970s and the property boom is a case in point. Ease of credit can result in speculative booms and 'over-the-top' borrowing by developers.

Statistics of capital issues and redemptions in the UK by property companies and of bank loans and advances to property companies are compiled quarterly by the Bank of England and published in *Financial Statistics*.

15.2.3 Information on the individual property

The list of all types of information relating to the physical, legal and financial attributes of individual properties and portfolios is very lengthy. Fortunately, most of these are, or should be, familiar to surveyors and most firms do keep copious records. It is therefore only necessary to identify briefly some of the less common aspects which do not arise in the day-to-day context but are, nevertheless, important.

Information pertaining to planning, land use and development control can be found in structure plans and planning reports. This information, although often taken for granted, is important for investment analyses of properties which are subject to changes in land use zoning of properties seeking planning permission. In some cases, similar information on surrounding properties is also needed.

Local information, such as traffic and pedestrian counts, new amenities in the area, market surveys, and so on, may also be essential raw material for investment analysis. Most of this information is usually used in research studies which support the analyses.

15.2.4 Summary

The information needs for property investment analysis, discussed above, are by no means a complete account. The discussion is, however, intended to fill the gap in the literature on property information. In the process, many references have been looked into, a lot of which are not formally published. The few quoted represent the limited and sporadic publications on property information. Undoubtedly, with more time, a more up-to-date and complete review can be achieved.

15.3 PROPERTY INFORMATION SOURCES

In the previous section, the information needs for property investment analysis have been identified. Each of these information elements can be presented in a very wide range of formats, especially with the aid of advanced technology. Taking just one important area of information, for example, property companies, the different kinds of sources which might be used include directories, annual reports, newspapers, periodicals and handbooks, and other agencies such as banks, investors' services and, of course, the Companies Registration Office.

Most reference guides to information sources are presented according to the types of source. There are a few which divide the sources into primary and secondary. Some examples of secondary sources are news and related media, statistical services, stockbrokers' reports and industry studies. Although the categorization of secondary sources is quite clear, the delineation of some of the primary sources is tenuous. For instance, whereas official statistics may be regarded as a primary source, some would treat it as secondary because it has been processed and hence manipulated. The controversy hinges on the definition of secondary sources, a generally accepted definition being any source of which the user is not personally responsible for collection and preparation 'from scratch'.

In terms of property information sources, there are very few true primary sources available. Most of these belong to private sector organizations, which treat them with confidentiality because of their commercial viability. The government, in the form of the Inland Revenue, and other local authorities also produce primary information. However, the forms in which the information eventually becomes available are often secondary, as, for example, official statistics from the Central Statistical Office. There is therefore an extensive but inefficient market for primary sources of property information. The sub-primary market of informal and anecdotal information is, however, fairly effective and is often relied upon as a source of vital information. There are also very few known tertiary sources, given the scarcity of property research in general.

The majority of property information sources are therefore secondary and are widely available but often poorly collected and publicized. In view of this, the discussion is best presented in terms of the various types of sources of information with examples to illustrate. However, the increasing importance and presence of electronic information needs to be emphasized. The electronic database, and the industry that evolves from it, is establishing itself as an unrivalled source of information because of its ability to provide information directly and almost instantaneously via online computers. One of the main characteristics of business and investment information is that it is transient and its value deteriorates rapidly in time. With printed information, there is considerable delay in the updating and this time lag can be costly, especially in the fast-moving world of the financial market, the proliferation of online services is a clear indication of the trend towards greater reliance on the computer as a medium of communication.

A parallel development to the fast-growing electronic information industry, in terms of the concept of selling information as a commodity, is third-party research or information services. Although this idea is not new, the recent formation of a number of property information firms is testimony to the increasing acceptance by surveyors of the use of these services.

The growth of these 'new' sources of information, however, does not imply the replacement of the traditional sources. Most of those in existence still constitute a very important part of research libraries and information centres. Therefore the following discussion on property information sources is perhaps best divided into these two categories (Fig. 15.2).

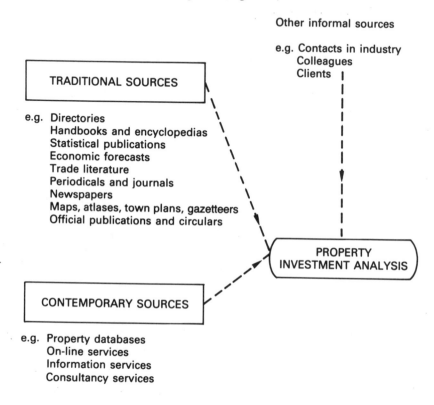

Fig. 15.2 Information sources for property investment analysis.

15.3.1 Traditional sources

Directories: Directories of various kinds are perhaps the most common form of information sources in the library, they provide a listing of individuals (e.g. 'who's who'), companies, societies or other organizations. Some deal with more specific sectors and may be concerned with a particular trade, industry or profession.

Handbooks, encyclopedias and dictionaries: These are essential references since they provide comprehensive treatment of specific subjects (handbooks) or a glossary of terms (encyclopedias) or definitions and meanings (diction-aries). There have been many property reference books published over the years, the most important of which are on US real estate, A–Z guides and glossaries (as a part of textbooks). Two relevant UK examples are the *Estates Gazette/Jones Lang Wootton Glossary of Property Terms* and the recent *Encyclopedia of Real Estate Terms* by Investcorp.

Statistical publications: Statistics are a basic resource for investment analysis and research. The importance of quantitative data for decision-making means that users need to be familiar with locating the major statistical sources. Fortunately, most statistical series are now publicly available with the development of the central government's role in the collection and dissemination of statistics on almost every conceivable human activity (the Central Statistical Office, CSO). Comparative statistical series are also made available by regional and international agencies, such as the EC, OECD and the United Nations. Special interests groups, such as trade associations, often produce detailed statistics for their own industry, and commercial publishers draw together statistical tables from these sources and enhance them by analysis and interpretation. In terms of presentation, statistics are often published in graphical formats, such as histograms and pie charts, as well as in traditional tabulation. Some of the major publications, for example, *Main Economic Indicators*, are already available in magnetic-tape format and it is likely that in future more will adopt such a form. There are also a number of guides and directories to the various statistical sources. The main ones are the *Guide to Official Statistics* (HMSO), *Business Statistic Index* (Headland Press) and *Sources of Statistics and Market Information*.

Economic forecasts: Economic forecasting is increasingly becoming an important source of information for investment analysis. Scores of main economic forecasts are now published regularly by official bodies, independent consultants, stockbrokers, academics, pressure groups and the clearing banks. Most are based on computer manipulation of economic models, although a growing number, such as the CBI's *Industrial Trends Survey*, are based on attitude surveys. In addition to tabular information, most forecasts carry expert analyses and commentaries. A recent development has been the emergence of surveys looking at specific industry sectors and geographical regions.

Trade literature: The term 'trade literature' comprises diverse publications. They may include company annual reports, house journals and product literature such as catalogues, data sheets and price lists. In the surveying world, the vast majority of firms are private partnerships and they are therefore not required to publish their annual reports in accordance with the rules of the Companies Act 1989. They invariably, however, produce brochures, journals, newsletters, market reports on different property sectors and research reports.

The main purpose of these is to inform and impress private and institutional clients with their services, experience and reputation. Despite this, most of this freely available literature (some are beginning to charge a fee to reflect the better standard of publication) does provide some important secondary information. The amount of useful information, of course, depends on whether it is a glossy piece for public relations or a serious report on research carried out. Most of the large, London-based surveying firms produce a whole range of trade literature on a fairly regular basis.

Periodicals and journals: In a fast-moving and rapidly developing world, periodicals are a particularly important source of information. The three main UK property periodicals are the *Estates Gazette* (*EG*), the *Chartered Surveyor Weekly* (*CSW*) and the *Estates Times* (*ET*). The former is a well-established source of property information with weekly coverage on every aspect of the property market. It contains up-to-date news of the property industry, movements of personnel, auction prices, law reports and commentaries, feature articles and regular focuses on specific topics. The *CSW* is a similar publication produced officially by the RICS and distributed to members. The *Estates Times*, in tabloid form, contains more feature articles but less coverage of all aspects of property news than the other two. There are other lesser-known and less frequently produced property periodicals, such as *Property Director* and *Property Business*, which may provide useful information. The other non-property periodicals which carry information on the economy with occasional property features include the *Economist*, *Business Week* and *Fortune*. As for journals, most of these are academically oriented. They are important for bringing together research and discussion which contribute to a better knowledge of the property market and its processes.

Newspapers: The information provided in daily newspapers is obviously even more topical than that in weekly or monthly periodicals. The *Financial Times* (*FT*) is the leading newspaper in providing extensive coverage of the complete spectrum of the financial markets. The Financial Times Indices are in fact the most widely followed indicators of the performance of the stock market and its various sectors. The *FT* also carries features on property once a week. Other major dailies, including *The Times*, *Guardian*, *Daily Telegraph* and the *Independent*, are also an important part of a research library because they normally report news before many of the other sources.

Maps, atlases, town plans, gazetteers: These are, again, essential sources of locational information which are indispensable to surveyors.

Official publications and circulars: Apart from the statistics produced by the CSO, there are other official publications which may be of use as information sources. These include legislative documents, annual reports and accounts of government departments, parliamentary debates and House of Common papers. Information on government issues and policies can be found in Green Papers and White Papers, Bills of Parliament and, finally, Acts of Parliament.

15.3.2 Databases, online systems and information services

Property databases

The distinction between this category and online services is principally one of the convenience in identification. Whereas online services are, as the name implies, accessed by online searching of the computer databases, the following property databases do not yet have online searching facilities.

The biggest property database at present is the Investment Property Databank (IPD). IPD is an independent research body sponsored by six firms of chartered surveyors. The sponsors provide the information for the database which in return acts as an important source of information to them. The database holds records of 7500 properties belonging to more than a hundred pension funds, insurance companies, property unit trusts and traditional institutes. IPD, as a research body, uses the database to examine trends in institutional investment, marketable size and market structure; to provide accurate market information about rents, vacancy and purchase prices; and to construct performance measurement indices. The results of these activities are published yearly and are available to subscribers. They do not, however, have access to the database as do the sponsors.

The Applied Property Research (APR) database, provides a specialist computerized information and consultancy service on the Central London Office Market, including the Docklands. The database has four main categories: information about new buildings by location, size, type and timing; information about companies involved in development and construction; information about demand; sector analysis, identifying and recording major demand trends and a comprehensive record on major occupiers by business sector.

Databases of the major surveying firms are also part of this category, although they only serve their own surveyors. From these confidential databases, the firms produce research and provide services to clients. At present, most of these publications are distributed free as part of public relations, but some have started to charge a fee, especially to those who are not their clients. This may start a trend towards an information market which brings together buyers and sellers of information from property databases.

Online services

Online access and search facilities are available through various telecommunication channels. The most common is the searching of databases by means of a computer terminal. The terminal is coupled through a telephone into a communications network which connects to a large, time-sharing computer in another city, country or continent. This computer provides access to one or more files stored on magnetic disks, and users are able to type in search commands and receive immediate responses in an interactive way. The two main parties concerned with this provision of information are the database and their producers and the host systems which convert information provided by the database producers into machine-readable records and make it available online. There are two main types of databases, bibliographic and source data. Bibliographic online searching, the forerunner of online information services, allows users to interrogate bibliographic databases and retrieve citations or abstracts of books, articles, reports, etc. Source databases allows retrieval of information which can be numeric, textual-numeric and full text. There is an increasing convergence of both types of databases, especially in the business information field, as numeric and full-text databases are overtaking bibliographic databases in importance. The users of online searching normally pay a subscription charge to gain access to the host system or database, as well as the computer processing time. There are also the indirect costs of user manuals, equipment and training.

The online information services are usually referred to by the generic term for television-based services, videotext. The two broad categories are teletext (broadcast over the airwaves), for example, Ceefax, Oracle, and Viewdata. Viewdata services are provided via the public telephone system through which the databases are connected to the users who receive the information on television screens. The users must have a suitable modem and modified television receiver, together with a simplified keyboard to communicate with the central computer. Prestel is the UK public viewdata brand name.

The number of online information services has increased dramatically in the last few years. This has meant that property investment analysis, as with other financial analyses, can no longer afford to ignore these services as a major source of information.

Most of these online services are relevant in the investment context and some do provide specific coverage on the property market. An example of a specialist property database for online information is FOCUS. The kind of property intelligence covered by FOCUS includes ownership, deals, company portfolios, rent review dates, planning and development, specific properties and properties available. It is a central online database providing a single reference point for published information about commercial property in the UK. By subscribing to this service, a user can conveniently access information which is called from a diverse range of sources. In this respect, it is increasingly being recognized as a helpful backup to, and verification of, the existing sources of information.

The growth of FOCUS, and its spin-off, Town Focus (providing demographic and socio-economic information on towns in the UK), indicates that there is still scope for further development of specialist property online services. The problem that remains is the fundamental issue of the primary source of data, especially with regard to accuracy and reliability. Unless surveyors can accept and understand information as a resource and not as a subjective privilege, the process of treating information as a commodity will be a slow one.

Other market information services

Market information services in the form of consultancy reports and third-party research are also recent developments in the surveying world. Applied Property Research (APR) and Property Market Analysis (PMA) are two such property research firms which provide these services. Both have databases from which information is accessed for carrying out research and analysis into specific areas of the property market. APR, as mentioned earlier, concentrates both on selling information from its databases and on consultancy. PMA, on the other hand, develops information systems in the course of conducting research for clients. PROMIS, for instance, is PMA's regular survey of the forty largest regional and sub-regional centres outside London, providing information on trends in the local economy and in the local office, industrial and retail markets, together with comparative summaries of regional trends. Other providers of research and consultancy include CACI, which has a property group for specialist research, and stockbrokers, such as Kleinwort Grievson, Barclays de Zoete Wedd and Wood McKenzie.

There are also information services, such as the Planning Exchange, specializing in a specific area of property information which are now entering the scene. The monthly information service of the Planning Exchange, for example, provides concise information on innovative urban development initiatives and reviews current policy and practices. This service can give developers insight into how developments have been facilitated in urban areas where there is little or no mainstream property investment.

The Property Developer Direct provides up-to-date details of office, residential, retail and industrial sites and properties available for development and investment. Subscribers will receive a directory of this information every fortnight.

The level of market information services in the property industry is, however, nowhere near that of the other financial markets. The corollary of this is that there is scope for further developments in this area. It can only be to the interest of property investment analysis that more market information services are available as a source of information.

Summary

The sources discussed above are of the most importance in finding information for property investment analysis. Inevitably there are a few minor sources which might have been overlooked here but have become established in a particular surveying firm over the years. These are peculiar to the practices of the organizations and do not usually exist in others.

There are also the informal sources of contacts in the industry and the proverbial 'grapevine' which are undeniably valuable sources of information for the surveyor. Together with his own knowledge of the market, these are often quoted in response to the question, 'Where do you get your information from?'

The list of sources discussed here is intended to be descriptive rather than prescriptive. In describing the sources the issues of quality in terms of accuracy, timeliness, ease of use, time-saving ability and promptness in delivery are not discussed. To do so, the scope of this research would need to be expanded even wider and a host of prescriptive measures would need to be addressed.

REFERENCES

Bakewell, K. G. B. (1987) *Business Information and the Public Library*, Gower, Aldershot.

Coghill, R. (1976) Financial information and where to get it, *Accountancy*, April, 46–7.

Edwards, M. (1987) Perspectives on the land and building development process, *Planning, Practice and Research*, Sept., 3, 20–22.

Fleming, S. (1985) The small organisation model, *Information Management: Strategics into Action*, (ed. B. Cronin) ASLIB, London.

Fraser, W. (1984) *Principles of Property Investment and Pricing*, Macmillan, Basingstoke.

Hager, D. P. and Lord, D. J. (1985) The property market, property valuations and property performance measurement, *Journal of the Institute of Actuaries*, June, **112**(1), 19–49.

Jaffe, A. J. and Sirmans, C. F. (1981) Improving real estate investment analysis, *The Appraisal Journal*, Jan., 85–94.

Jones Lang Wootton Property Contacts (1988), (4th ed.), Newdata Publishing, London.

Maunder, (ed.) (1978) *Reviews of UK Statistical Sources*, VIII, Pergamon Press, Maidenhead.

Salway, F. (1986) *Depreciation of Commercial Property*, CEM, Reading.

The processing and management of information

16.1 INFORMATION IN DECISION-MAKING

> The relationship between availability of information and decisions is basic to the economic theories of decision making. (Bruns)

Information plays an essential role in the functioning of purposeful, decision-making entities and the flow of information amongst such entities is basic to all social system. It is a raw material, an intermediate product and the finished goods of a decision-making process. Any form of decision-making requires information. Whenever a decision is being made, information concerning all the elements of the decision will shape the decision-making process and ultimately the choice itself. The level and amount of information which will be needed may vary for differing decisions but the need for information, whether explicitly or implicitly expressed, will always be present. Information processing is therefore an integral part of decision-making.

Decision analysis or the science of decision-making is concerned with the study of factors or information which affect the decision problem, as well as the evolvement of techniques which attempt to clarify and analyse the problem in such a way as to increase the chances of attaining consistent and acceptable results. To improve the quality of decision-making, the three approaches are: first, better understanding of theory of decision-making under uncertainty; secondly, application of techniques and systems which will help to improve the process; and thirdly, improve the quality and increase the quantity of information.

However, it is argued that techniques, no matter how sophisticated they are, cannot minimize the uncertain elements of the future, nor can they guarantee a precise knowledge of the outcomes. Information processing especially of relevant and up-to-date information can, however, improve the decision-maker's perception of the problem which will ultimately provide a better chance of attaining the decision-maker's objectives and adhering to them. Kent (1971), for example, states that: 'the quality of decisions depends fundamentally on the problem-solving capability of the decision-maker, but initially on the quality and relevance of information brought to bear on the problems. Information can make the difference between a decision and a guess, which can mean success or failure.

Janis and Mann (1977) provide a list of seven 'ideal' procedural criteria which they call 'vigilant information processing'. Accordingly, the decision-maker to the best of his or her ability, (and within his or her information processing capabilities) should:

(a) thoroughly canvass a wide range of alternative courses of action;
(b) survey the full range of objectives to be fulfilled and the values implicated by the choice;
(c) carefully weigh costs and risks of negative consequences and benefits of positive results of each alternative;
(d) intensively search for new information;
(e) correctly assimilate and account for any new information or expert judgement;
(f) re-examine all consequences before making a final choice;
(g) make detailed provisions for implementing or executing the chosen course of action, with special attention to contingency plans that might be required.

Although these criteria by and large correspond to the stages involved in any rational decision-making process, they do highlight the importance of the role of both existing and new information.

16.2 INFORMATION IN MANAGEMENT

Information is the manager's main tool, indeed the manager's 'capital' and it is he who must decide what information he needs and how to use it. (Drucker, 1987)

Besides being an integral part of the decision-making process, information is also a vital aspect of management. Strictly speaking, it is difficult to distinguish management functions from decision-making, as the core of management is about decision-making. However, by adopting the perspective of management, the scope can be further enlarged for the appreciation of the importance of information.

Simon (1979) has analysed organizations on the theory that organizational members are complex information processors and that organizations are basically large decision-making units. His major thesis is that the foundation of the organization is the flow and rational application of information (knowledge) to problems confronting the organization. This view stresses the importance of acquiring, processing and disseminating information to improve rational decision-making and to meet organizational goals.

Traditionally, information was not recognized as a resource in the way that most economic theories recognize land, labour and capital as resources. Information was only incidental to management which included such functions as planning, controlling, co-ordinating, organizing and leadership. The classic model of economic production assumed that an enterprise ought to have most of its resources in production, that is in human and capital resources (operators) that concentrate fully on the delivery of physical goods to customers. Underlying this concept of organization is the idea that the information processing needs of the firm are secondary and can be satisfied by management or, when necessary, by employing specialists.

However, as the industrial society gave way to the service society, the idea of communication and subsequently the concept of information itself began to surface in studies of behavioural science and management. As global competition increased and government intervention intensified, and as the pace of technology accelerated, enterprise became much more complex. Organizations began to employ large numbers of professional, administrative and technical workers in positions which were closely identified with management.

The increased information flow through this more complex structure of management eventually contributed to the recognition of information and its role in management. Some commentators assume that the main reason for this new-found interest in information was the development in computer technology. With the technical ability to capture, transmit, store, process and retrieve masses and text globally, attention is now turned to the manager's information needs. However, as stated previously, this is only true because the dominant reasons for that new-found interest in information are the changes which have taken place at the management level. Some of these changes which affect the need and therefore the role of information are: the increasing scale, concentration and diversity of business; the greater complexity in decisions; and the increasing external demands for information through increasing institutional regulations. A simplified diagram of the information flow in a typical present day firm is given in Fig. 16.1.

From these developments, it can be seen that information is indeed the basis of 'significant power'. The larger firms are now regarding information as a critical corporate resource just like conventional human and capital resources. It has become managements' principal need, and managing the information resource their most crucial activity.

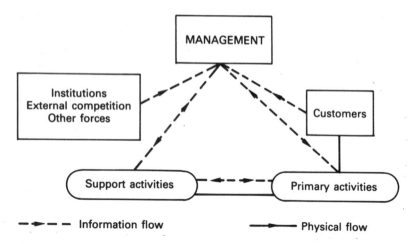

Fig. 16.1   Information flow within a typical present day firm.

16.3 THE INFORMATION TECHNOLOGY REVOLUTION

The increasingly recognized fact that information is important is not due to a sudden realization, but rather it is the result of the evolving technology that supports it. The role of information has not changed, but what was missing before the current technological era was the existence of advanced electronic equipment in the form of computers and other telecommunication technologies. Indeed, it was only about a decade ago that the prevailing public image of Information Technology (IT) was one of mysterious and highly expensive machinery which had to be discussed in esoteric jargon.

Growth in the technology, much of which barely existed twenty-five years ago, has been rapid and continuous in the past decade. Since its first business application in the mid-1950s, the use of computer technology in all areas of organization and business has expanded dramatically. The picture confronting the firm is constantly evolving, and few weeks go by without announcement of new products, major technological advances, improved cost performance ratios and myriad allied services. Fired by parallel developments in marketing techniques, the typical business is being bombarded by a host of technologies, many of which bear the most inscrutable names, mnemonics or just plain numbers that convey little or nothing to the uninitiated. Similarly, the range of activities to which computers and cognate technologies can be applied is vast and there are few aspects of the working and leisure lives of the individual which are not influenced in some measure by IT.

The 'revolution' in IT has been the subject of numerous publications and the topic of many conferences and seminars. Assertions such as: 'The Information Revolution is sweeping through our economy. No company can escape its effects' are frequently made in findings of surveys and such studies of the economic environment. IT has also pervaded every form of human activity to the extent that everyone will be affected by it at one time or other. The result of the IT revolution is that its penetration into the fabric of society has reached far and wide and is, by now, almost certainly irreversible.

16.4 DEFINITIONS OF INFORMATION TECHNOLOGY

The definitions of IT are as wide-ranging as the range of products based on the microchip. A typical definition is that it has come 'broadly to encompass the information that business creates and uses as well as a wide spectrum of increasingly convergent and linked technologies that process the information'. In addition to the ubiquitous computer which is the key element in the information revolution, data recognition equipment, communication technologies, factory automation and other hardware and services are involved. Some definitions are perhaps more precise and relevant, as, for example, 'the science of information handling, particularly by computers, used to support the communication of knowledge in technical, economic and social fields' or simply 'the application of science to the collection, storage, processing and dissemination of information'. Some are more nebulous and all-embracing; for example, UNESCO defines information technology as the 'scientific, technological and engineering disciplines and the management techniques used in information handling and processing; their applications; computers and their interaction with men and machines; and associated social, economic and cultural matters'. This is perhaps superfluous in the context of this study, as the main concern with Information Technology is in the aspect of gathering, processing, interpretation and communication of information by means of computing technology.

Whatever the definitions, it is apparent that technology goes hand in glove with the work carried out in the service industry. It is therefore important to understand the information tool which supports information work and to see it as an indispensable resource. In fact possibly the greatest barrier to the application is of how it works and of what it is capable of achieving, and apathy towards understanding, especially the significance of its impact.

16.5 APPLICATIONS OF INFORMATION TECHNOLOGY

The definitions of IT given above broadly embrace all kinds of computing technologies which assist the processes of information handling. These are the five stages of input, processing, storage, transmission and output. Some or all of these five stages are built into various systems, which are made up of equipment (hardware) and, most important, logical programs (software). These systems can be individual microcomputers dealing with all the stages locally on a small scale, or large networks based on mainframe computers.

Apart from the computer which is the central element of the current revolution in IT, there are other applications which need to be mentioned. These applications are discussed briefly, but their influence on information is particularly stressed. The discussion also bears in mind their relevance to the property world, the specific applications of which are discussed later.

The development of management information systems in the past decade has led to the creation of a separate discipline on that subject area. Some basic concepts are explored and their relationships with information are emphasized. Another application of growing interest and importance is that of artificial intelligence, and specifically, the creation of expert systems. The advancement in technology has enabled such application in professional services, as well as other forms of robotics and automation, which are relevant to the manufacturing industries. The explosion of electronic, on-line information services also needs to be reviewed with regard to its impact and implications to information handling.

16.5.1 Computer applications

The name 'computer' was originally given to machines because computation was the only significant task undertaken. This name somewhat obscures the fact that they are capable of much greater generality and perhaps a more embracing, generic name such as 'information machine' would be more appropriate. However, the term 'computer' itself is significantly the only thing that has not changed since the first general-purpose machine was built in the mid-1940s.

In terms of the rate of advancement, computer circuitry has continued to improve unabated at about 20% per year in the price–performance ratio. The physical configuration has also changed dramatically as a result of advancement in general technology. Input, storage and output devices have become sophisticated and powerful. The growth of the hardware industry has also given rise to the proliferation of software, especially in database management systems and application development tools. These software packages are easy to use and are enhanced and tailored to specific needs. The advance in telecommunications, which is the electronic transmission of data from one location to another, has enabled services such as an integrated services digital network and a local area network. These developments are particularly relevant to information-based services.

Management information systems

Any organization can be viewed as a total system comprising three sub-systems, namely operations, management and information. The operations sub-system includes all people, activities and material flow related to performing the primary functions, whilst the management sub-system includes the people and activities which determine the sub-system. The information sub-system is then the aggregation of people, machines and activities that gather and process data to meet the information requirements of the organization (See Fig. 16.2).

The advancement in computer technology, described earlier, has brought parallel developments in information systems application. **Management information systems** (MIS) have become a separate discipline of study. Various concepts and applications have resulted directly from the improvement in technology and the advent of the Information Age. Terms such as 'Decision Support System' and 'Information and Decision System' are used to identify specific concepts and objectives. With reference to the literature, there is also a proliferation of articles and publications on a host of issues relating to MIS. For example, the question of evaluation, design and implementation has been the subject of numerous research studies and textbooks. The aim of this discussion, however, is to examine two aspects of information systems development which have a direct influence upon property information.

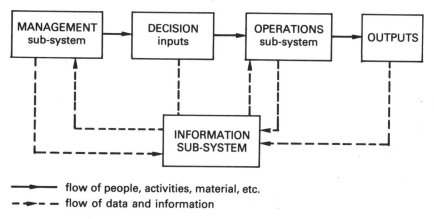

flow of people, activities, material, etc.
flow of data and information

Fig. 16.2 Information flow within an organization.

Although there is no consensus on the definition of the term 'information system', it refers by and large to a computer-based information processing system which supports the operations, management, and decision-making functions of an organization. In the early development of information systems in the 1960s, the emphasis was on the management of an increasing amount of data. The proper storage, retrieval and delivery of information was the primary objective. The development in computer technology during this period was similarly concerned with large storage capacity and speed of processing. The progress in technology therefore helped the growth of information systems which became capable of storing and processing vast amounts of data. Many of the increasing number of mainframe computers in the 1970s were used to support the information systems of government bodies and large private organizations.

However, whilst the technology continued to improve, the emphasis shifted to the quality of the information produced from the systems. As information systems were made up of the computer (data processing) as well as the information (people, ideas and activities) system, the attention is now focused on the latter. The concepts of management and decision support have become more important.

One of the key concepts of the current MIS development is that of 'value-addedness'. Taylor (1982) defines information systems as a series of formal processes by which the potential of the information is enhanced or value is added. Whilst most of the conventional information systems are concentrated on internal aspects, in particular the content and the technology of the systems, Taylor suggests that their usefulness can be improved by adopting a value-added approach which can provide a powerful organizing concept. It requires an understanding of systems as value-adding processes and the information environment from which tasks and problems arise requiring information for resolution. In contrast to the conventional definition, information systems are described as activities which are formally designed processes that select, acquire, organize, store, retrieve, display, analyse and interpret messages. Through these processes value is added to the messages.

Property information systems

There are two basic applications of information systems within the context of the property industry. The first concerns the use of the computer to store various data with regard to property. The second is the increasing application of the MIS/database management concept to property management.

The earliest example of a property database is 'Domesday Book'. From this historical point, land information of one kind or other has been recorded and stored over the years. Both national and local authorities have set up land information systems for planning, housing, environment land titles and valuation. The importance of these land information systems has been increased by the advancement in IT. This is summed up in a report on handling geographic information: 'Large sums of money are spent by government, commerce and industry . . . in collecting and using it [geographic information]. Much human activity depends upon the effective handling of such information . . . but it is the ability of computer systems to integrate these functions . . . that makes them such potentially powerful tools.' Whilst this is of no direct relevance to investment·analysis, the suggestion of the committee to make available more property and socioeconomic data, such as is held by the Land Registry and Valuation Office, is of some significance.

However, the growing importance of the role of these land information systems can only be realized if the benefits of IT are harnessed. The traditional land systems have been criticized for merely 'automating the inefficiencies of the past'. This has resulted in a growing body of literature on developments in land information systems.

As for information systems in property management, there are two forms of development. The first is the specially designed program based on a mini- or mainframe computer. This is widely used in local authorities, and the larger surveying organizations. The second is the standard database management packages which are commercially available. Users can either commission a specialized computer firm or develop an in-house program based on standard software. In both cases, the concepts of decision support are increasingly incorporated into the design. The proliferation of available systems has also provided alternatives to potential users. This clearly allows more room for proper assessment and planning.

16.5.2 Expert systems

An important development arising largely from artificial intelligence research is the creation of an expert system. It is defined as a computing system which embodies organized knowledge concerning some specific area of human expertise to perform as a skilful and cost-effective consultant. It is thus a high-performance, special-purpose system which is designed to capture the skill of an expert consultant such as a doctor.

The earliest acknowledged expert system, DENDRAL, dates back to 1965. The term became more popularly used only in the late 1970s. Although the definition places emphasis on performance, it is the methodology which provides the identification. Typically, expert systems make use of heuristic rules which approximate the expert's knowledge of the domain in question. These rules are acquired from subject experts and refined in the light of experience. The skill involved in setting up an expert system is often referred to as 'knowledge engineering'. Although expert systems can vary considerably from one another in terms of design and capabilities, not least because the term is not yet precisely defined, most systems have many features in common. These include a knowledge base, an inference mechanism which manipulates the knowledge base to form inferences, and the provision of an explanation facility.

The increasing use of expert systems in various professional fields has brought about debate concerning their practicality and cost-effectiveness. There are, however, some positive applications which do not claim to be able to exhibit total human expertise but are user-friendly systems that help decision-making. Practical uses of expert systems have occurred in the areas of medical diagnosis, geological exploration, insurance underwriting, computer system configuration, machine repair and maintenance and financial and marketing analysis and planning.

The increasing applicability of expert systems to industry and commerce has also spread to the surveying profession. However, this introduction of expert systems valuation is still largely being researched and discussed academically. The practical application has yet to gain widespread commercial acceptance.

16.5.3 On-line services

Searching for information in computer databases is a very different process from the search in a library of printed material. It takes the form of a dialogue between the searcher and the database or electronic library and is, therefore, an interactive process. The attraction of this has led to a phenomenal increase in the use of electronic information services.

The computer was applied, in the first instance, to automating existing indexes and abstract journals. Reference databases were then joined by primary-source databases. This development has given rise to the burgeoning electronic information industry. There are four key issues which have influenced this rapid growth. First, the increasing acceptance of information as a tradeable commodity, despite the externality of it being a 'free' good. The willingness of the commercial world to pay for information, particularly financial services, has led to the creation of business enterprises dealing solely with information. Secondly, the advent of relatively cheap and powerful telecommunication and computer facilities has helped the sale of information not only to businesses, but also to lay people in their homes. Thirdly, the further impact of new technology in the form of artificial intelligence is bound to help increase the use of on-line services as intelligent terminals acquire the ability to assess electronic databases. Finally, the concept of value-addedness in information broking is increasingly applied to on-line services where the basic information is repackaged and tailored to specific users.

There are other technological innovations, such as office automation, which also affect the offices of general practice surveying firms. The convergence of office and computer technology has transformed Dickensian practices into a modern environment where ergonomics is the key issue. As the pace of technological advancement continues unabated, it is important to note that what has been described may not be relevant in the future.

16.5.4 Conclusions

The application of IT in the surveying profession, described above, reflect the availability of opportunities to incorporate IT in its provision of services. Although the level of application has often been said to be comparatively low, *vis-à-vis* the USA and other professions, the growing number of technology-related publications indicate the recent surge in awareness.

The development of property information systems is contributory to the overall collection and processing of data, particularly those designed for decision support.

The creation of expert systems, although still in its infancy, has particular significance for this research. In the development of the expert system for residential valuation, for example, the creation of the knowledge base relies on the distillation of judgements with regard to information.

REFERENCES

Adair, A. and McGreal, S. (1987) The application of multiple regression analysis in property valuation, *Journal of Valuation*, **6**, 57–67.

Baum, A. (1984) The all risk yield: exposing the implicit, *Journal of Valuation*, **2**, 229–37.

Bernal, J.D. (1957) The supply of information to the scientist: some problems of the present day, *Journal of Documentation*, **13**(4), 198.

Brittain, J.M. (1970) *Information and Its Users*, Bath University Press, Bath.

Crosby, N. (1986) The application of equated yield and real value approaches to market Vol. I, *Journal of Valuation*, **4**(2), 158–69.

Crosby, N. (1988) An analysis of property market indices with emphasis on shop rent change, *Land Development Studies*, **5**, 145–77.

Derr, R.L. (1983) A conceptual analysis of information need, *Information Processing and Management*, **19**(5), 273.

Dervin, B. (1977) Useful theory for librarianship: communication not information, *Drexel Library Quarterly*, **13**(3), 16–32.

Drucker, P.F. (1987) The knowledge economy, *Evolution of an Information Society*, (ed. A.E. Cawkwell), ASLIB, London, p. 76.

Frazer, W. (1986) The target return on UK property investments, *Journal of Valuation*, **4**(2), 119.

Hager, D.P. and Lord, D.J. (1985) The property market, property valuations and property performance measurement, *Journal of the Institute of Actuaries*, June, **112**(1), 19–49.

Hoag, J.W. (1980) Towards indices of real estates value and return, *Journal of Finance*, May, **35**(2), 569–80.

Janis, I. and Mann, L. (1977) *Decision-making: a psychological analysis of conflict, choice and committment*, Free Press, New York.

Kent, A (1971) *Information Analysis and Retrieval*, Becker & Hayes, New York, p.2.

Leigh, D. (1986) Investment profiles, *Journal of Valuation*, **4**(3), 291–9.

Rouse, W.B. and Rouse, S.H. (1984) Human information seeking and design of information systems, *Information Processing and Management*, **20**(1–2), 129.

Simon, H.A. (1979) Rational decision-making in economic and behavioural science, *American Economic Review*, June, 253–83.

Taylor, R.S. (1982) Value-added processes in the information life cycle, *Journal of the American Society for Information Science*, **33**(5), 341–6.

Property Decisions in Practice: Examples

17.1 INTRODUCTION

This chapter explores a small sample of property-related decision-making situations as they are approached, analysed and resolved in practice. Section 17.2 is the work of Brigitte Hague of Norwich Union Real Estate Managers. This section contains an example of producing a Report to the Trustees, based on the performance appraisal of a portfolio of property assets. The presentation of this example is normative – i.e. drawing attention to those aspects which should be (and should be expected to be) included in a portfolio report. In the example the IPD Annual Index is used as the benchmark in a manner considered appropriate by the author. The author's views and interpretation of IPD indices are not necessarily those of the Investment Property Databank Limited.

Sections 17.3–17.5 are provided by Ann Colborne, of Bristol Polytechnic and member of the Investment Surveyors' Forum. These sections contain various examples and possible solutions to decision problems associated with ownership vs leasing, taxation, financing and accounting matters.

17.2 PORTFOLIO PERFORMANCE MEASUREMENT AND MANAGEMENT

17.2.1 The brief

The manager of a property investment portfolio of an investment trust is required to produce a Report to the Trustees. The Report must concentrate on the following aspects:

1. Measurement, analysis and explanation of the performance of the portfolio to the trustees.
2. Outline proposals to re-structure the portfolio to improve its future performance.
3. Justification of any further expenditure on the property assets in the portfolio and proposals for new acquisitions.

The example portfolio used to illustrate an explanation of portfolio performance and the proposals for restructuring to improve future performance in order to justify further expenditure on property assets is drawn from a selection of properties held as one portfolio.

Any explanation of portfolio performance and future strategy would be given in report format and therefore it is proposed that the question is addressed with the aid of a skeleton report format, giving an explanation of the requirements under each section. The portfolio structure is as given in Table 17.1.

Table 17.1 Portfolio ABC – portfolio structure and distribution by market value, (%) 1990

	Retail %	Office %	Industrial %	Other %	Total %
City		17.2	0.1		17.3
West End	1.5	10.1		0.1	11.7
Holborn/mid town		8.1			8.1
Other London	9.0	6.0	2.0	0.9	17.9
Total of London	10.5	41.4	2.1	1.0	55.0
East Anglia	4.2	2.5	0.5	0.1	7.3
South-east	8.3	6.4	3.1		17.8
South-west	4.1	0.5	0.3		4.9
Midlands	3.5	2.6	0.5	0.1	6.7
North	3.0	2.0	0.4		5.4
Scotland	0.9	0.7	0.1		1.7
Other	1.0	0.1	0.0	0.1	1.2
Total	35.5	56.2	7.0	1.3	100.0

Portfolio ABC performance and strategy, 1991

(1) Introduction
This should summarize the purpose of the report, i.e.:

- To explain 1990 performance.
- To identify means of improving future performance.
- To provide justification for continued commitment to new investment into property.

(2) Client objectives
It is useful to restate the client's objectives at the beginning of the report, highlighting the role of property, for example:

- Property expected to outperform fixed-interest investments long-term – useful where the client has long-term commitments.
- Property has good diversification qualities – useful where the client is keen on asset diversification and risks must be spread.
- Specialist area which increases uniqueness of the fund – useful where the client is keen to be seen to be different.

This section should also contain a sub-section restating the real estate strategy currently in place in an attempt to achieve the client's objectives. For example, the following comments could be made in respect of the example Portfolio ABC:

- The portfolio is required to outperform the IPD index on an annual basis.
- Previous strategy has been to invest in real estate which offers a good initial yield and steady rental growth over a 25-year period.
- Investment has been mainly through developments and funding as these have provided a higher initial yield than on ready-made properties, particularly in relation to large lot sizes.

(3) Performance results and analysis, 1990

This section should clearly and concisely show the performance of the portfolio. The use of tables can be most helpful in providing a quick overview and serve as important reference points.

The first requirement is to show the returns achieved during the year. The returns of Portfolio ABC are shown in Table 17.2. Figures obtained must then be compared with the objective of outperforming IPD as illustrated in Table 17.3.

Table 17.2 Portfolio ABC – total return matrix, 1990

	Retail	Office	Industrial	Other	Total
City		– 21.5	1.0		– 21.4
West End	– 8.9	– 13.2		2.0	– 12.5
Holborn/mid-town		– 18.1			– 18.1
Other London	– 12.5	– 10.1	2.1	5.0	– 9.1
Total London	– 12.0	– 17.3	2.0	4.7	– 15.1
East Anglia	6.2	4.3	4.2	3.7	5.4
South-east	– 3.1	– 5.6	3.9		– 2.9
South-west	5.7	2.1	5.6		5.3
Midlands	4.1	3.0	4.1	5.6	3.7
North	3.1	4.1	4.0		3.5
Scotland	2.0	5.0	5.3		3.4
Other	3.5	3.1	.	6.2	3.7
Total	– 2.1	– 12.6	3.5	4.8	– 7.5

Clearly, the major performance objective has not been met, and on current projections it is anticipated that the portfolio will under-perform IPD in the immediate future, given the current portfolio structure. It is therefore useful to examine the current situation in comparison with IPD sector and locational weightings, as shown in Tables 17.4 and 17.5.

Table 17.3 Portfolio ABC – performance objective analysis

| | Forecast | | | Total return % | | | | | | | | 10-year average |
	1989 %	1990 %	1991 %	1992 %	1993 %	1994 %	1995 %	1996 %	1997 %	1998 %	1999 %	total return %
ABC	16.0	−7.5	−11.2	9.9	16.5	15.0	15.2	13.3	13.1	13.5	12.1	8.6
IPD	14.8	−5.2	−10.2	10.1	16.0	14.1	15.0	12.8	13.2	13.9	11.7	8.7
ABC Outperformance	1.2	−2.3	−0.7	−0.2	0.5	0.9	0.2	0.5	−0.1	−0.4	0.4	−0.1

Table 17.4 Portfolio ABC – portfolio structure in comparison with IPD, analysed by market value and use

		Actual, 1990	
Class	ABC %	IPD %	Load ratio
Retail:			
Single units	15.6	20.4	76
Shopping centres	19.9	14.1	141
Office	56.2	53.3	105
Industrial	7.0	12.2	57
Other	1.3	0	–

Load ratio above 100 indicates ABC is overweight ⎫
Load ratio below 100 indicates ABV is underweight ⎬ in comparison with IPD's current weightings

Note:
The load ratio indicates how ABC's portfolio currently differs from the market average.

Table 17.5 Portfolio ABC – portfolio structure in comparison with IPD, analysed by market value and location

	Actual 1990		
Location	ABC %	IPD %	Load ratio
City of London	17.3	14.5	119
Rest of London	37.7	36.7	103
East Anglia	7.3	2.5	292
South-east	17.8	21.6	87
South-west	4.9	4.4	111
Midlands	6.7	6.5	103
Rest of UK	8.3	13.8	60

Clearly, the structural analysis shows a wide variation from the IPD sector weightings – these variations should be highlighted in the results report as shown above.

(4) Other investment markets and economic climate
This section should provide:

- A brief overview of the returns obtained from other investment media over the performance period.
- A brief résumé of current market economic factors (e.g. movements in base rates and their possible effects).
- Forecasts in respect of key economic factors (e.g. inflation, consumer spending and manufacturing output) as these will act as indicators to the key activity areas.
- A general résumé of the property market by use sector. This should be tailored towards the proposed new strategy recommendations.

Appropriate comments in relation to Portfolio ABC would be as follows:

- In considering the property market it is evident the investment performance could actually deteriorate further before improvements are seen. Although the rate of capital depreciation appears to be slowing for offices and shops, the slump is now beginning to affect the industrial sector. Whilst the slump in performance has been the result of a rise in property yields, rental growth has held up fairly well, although this is now beginning to change. If recession is deep-seated, occupier demand will weaken across all sectors. Thus falling rents could well take over where rising yields left off.

- The property market is typical of markets where supply cannot quickly respond to demand. The result for property is a cycle which moves continually between excess supply and excess demand. The implication of this is that the market will move from weakness to strength over the next five years as the current supply/demand imbalance is corrected.

- Considering the individual sectors, the following points should be noted:

Retail – deterioration in performance has slowed, probably indicating the bottom of the cycle. Demand for shop units remains weak, but with yields at very attractive levels limited buying could be seen in the near future.

Offices – the slump has been later than for retail, but more rapid. The deterioration has been inexorably linked with over supply problems in the City and this situation is unlikely to be resolved in the next two to three years. With yields expected to move out further and rental growth expected to decline, offices will shortly replace retail at the bottom of the performance league.

Industrial – this remains the best-performing sector, even though capital growth has entered negative territory. Given the vulnerability of manufacturing, the expectation is that yields will rise substantially. With rising unemployment, prospects for 1991 appear bleak but are more encouraging thereafter.

Conclusions – looking to the next five years, the current belief is that industrial rental growth will be strongest, followed by retail. Office rental growth is expected to be the lowest, the office sector suffering from both heavy oversupply and a reduction in the growth of financial and business services. From past experience, property market slumps tend to be protracted and hence recovery within the property market in 1993–94.

(5) Structure strategy and recommendations

This section should clearly outline any proposed re-structuring and proposals for more active management, including debt collection, refurbishment or re-development. The strength of the current assets should be highlighted and active management policies clearly defined.

In relation to Portfolio ABC, the major objective will be to show how the portfolio may be restructured to improve performance. A suggested approach could be as follows.

Analysis of the portfolio data against the IPD benchmark has illustrated that the portfolio is substantially different in structure in IPD. Whilst this has produced excess performance in past years from some sectors and regions, this imbalance is a major hurdle to future performance prospects.

In view of the inevitable capital restrictions in the current market, it is not recommended that the performance downside is addressed by counter-cyclical investment and thus it is necessary to reduce exposure to those sectors and regions with the worst performance prospects. Simultaneously, there will need to be new investment in key growth areas. It is important to stress that it is not the objective of the fund to match IPD exactly. Excess performance must be generated by moving into sectors and regions of greatest opportunity.

Developments in hand should be continued but new schemes should be of a much smaller lot size.

By projecting yields and rental growth, the returns for property by user and location has been forecast. This has produced the following buy (or hold) and sell recommendations, shown in Table 17.6.

Table 17.6 Portfolio ABC – recommended restructuring strategy

	Selection strategy ranking	ABC weighting (vs IPD)
Best buys		
Retail		
East Anglia	1	Over
Other south-east	2	Under
South-west	6	Over
Wales	8	Under
Industrial		
East Anglia	4	Under
London	5	Under
Other south-east	7	Under
North	9	Under
South-west	9	Under
Offices		
East Anglia	3	Over
South-west	11	Under
	Sales strategy ranking	ABC weighting (vs IPD)
Suggested sales		
Offices		
Central City	1	Over
Fringe City	2	Under
West End	3	Under

By undertaking further sector/regional research to identify those markets of greatest potential, it is anticipated that further investment into property can be justified and performance improved.

It is recommended that the proposed restructuring is phased over a three-year period such that the major sale of city offices does not prejudice the funds standing in the market to lead top disappointing capital receipts.

Predictive analyses show that by undertaking restructuring in line with the above – i.e. concentrating on good industrial purchases, offices in East Anglia and the south-west and major sales of city offices – it should be possible to achieve the stated objectives in relation to IPD by 1992.

Two further areas are also worthy of consideration:

(a) *Europe* – although further research is needed, exposure to the European property markets will increase diversification and provide opportunities to capitalize on property investment in potential growth areas.

(b) *Property shares* – recent analysis has shown a close correlation between property share movements and market movements generally, although timing differences can be observed with shares tending to anticipate market movements. It is recommended that the opportunities for using property shares to secure added performance from short-term market opportunities and from underweight sectors could be investigated. The flexibility and liquidity which would be introduced into the portfolio must, however, be balanced against the higher volatility levels experienced in property shares.

(6) Conclusions

This section should briefly summarize the key performance points and the key selling points which have to be implanted in the minds of the trustees in a bid to maintain their support for property investment.

The trustees should be reminded that all investment markets are suffering due to the economic climate and it can be useful to reiterate the role of property, as outlined under Section 17.2 as far as the investment trust is concerned.

As investment funds tighten and performance time horizons shorten, property investment justification becomes more crucial. The above clearly shows how many considerations and factors will have to be put to the trustees to maintain their support.

17.3 PROPERTY OWNERSHIP VS LEASING

The literature on this relates almost entirely to the purchase/lease of fixed assets with a limited life which can be acquired through finance leasing arrangements. Strangely the tradition of leasing property is so strong that property leasing is seen as an operational practice rather than one of finance. There is no logic in this. The working of the Landlord and Tenant Acts effectively secure long-term occupation to the tenant, who will pay a regular contractual rent to the landlord and must extend his lease at the end of the contractual tenancy (or seek other premises) to secure his long-term operational needs.

Current wisdom on this decision-making process recommends a net comparative approach based on the cost of financing by borrowing less the cost of financing by lease.

Gordon,[1] in 1974, produced this simplified version of the general formula:

$$\text{NPV (purchase)} - \text{NPV (lease)} = \frac{(1 - T)(L_t - M_t)}{(1 + i)}$$

where T = tax rate;

  L_t = lease payments at end of each period;

  M_t = payment of interest and principal on a loan;

   i = interest rate on risk-free cash flow (low debt yield);

   t = lease term.

Others have produced similarly based formulae. Brearley and Myers[2] suggest the use of an Adjusted Present Value to adjust for the taxation aspects of a leasing deal.

The major snag in applying this type of formula to property assets is its lack of adaptability to changes in lease rents and the creation of valuable profit rentals measured against rising property values where outright purchase is made. Consequently, the approach to decision-taking should be on a clear discounted cashflow basis providing the opportunity for changes in rental flow, preferred by Johnson and Llewellyn[3].

The outcome of analysis will usually show that the initial 'low' cost of leasing – sometimes as low as 4–6% of purchase price – a considerable saving on debt financing will be eroded over time as rents rise. Consequently, the statement: 'The lease vs buy decision is a financing one, not an investment decision' is true in the sense that a company should first make the decision to invest in a new asset on the basis of comparative investment appraisal – but it cannot assume that, however it is financed, the cost of the asset is measurable and fixed and that all parties to the prospective alternatives will use the same cost of capital or discount rates. Information on property lease decisions is almost entirely anecdotal and is not supported by generalized analysis which compares property leases to other leased asset decisions.

17.3.1 Real estate leases

Using the same intuitive approach, a difference can be seen in the way in which the two parties to a deal approach the problem of real estate leasing.

Example: Lease vs Buy Decisions

Land and buildings valued freehold at £3 million are required to fulfil an investment decision. The annual rental of such premises is £300 000 p.a. (note that the £300 000 is derived from market value considerations and is not a special price quoted to a particular lessee based on his corporate risk). The risk is based on the market security of real estate assets.

Ignoring for our purpose the taxation advantages, which disguise the more pertinent problem of interest rates: the cost to the company is the discounted cost of a rising cash flow over the next 20 years (the rental will rise in 5 year tranches), plus the cost of securing the residual asset to maintain long-term production and permanent occupation.

The value to the lessor is the capitalized value of the expected rental cash flows (i.e. the amount for which he could expect to sell the property – £3 million).

Valuation of rental outflows

Company cost of capital 15.6% net: expected rental growth 8% p.a.

Years	Cash flows	Net flows	PV Factor	NPV
1–5	300 000	195 000	3.3	643 500
6–10	440 800	286 520	1.6	458 432
11–15	643 570	418 320	0.77	322 106
16–20	939 600	610 740	0.37	225 974
21 (resid.)	13 718 250	13 718 250	0.05	685 912
				2 335 924

The additional taxation effects (i.e. the loss of any Capital Allowances) can also be incorporated.
The Writing-down Allowance of 25% p.a. on plant and machinery (say, 30% of the total cost) is £900 000 (reducing Balance).

Year	WDA	Tax credit @ 35%
1	225 000	78 750
2	168 750	59 062
3	126 562	44 300
4	94 922	33 222
5	71 191	24 917
6	53 393	18 688
7	40 045	14 016
8 (perp.)	120 137	42 048

Tax losses incorporated into the cash flow

Year	Cost	Tax loss	Net cost
1	195 000	78 750	273 750
2	195 000	59 062	254 062
3	195 000	44 300	239 300

Year	Cost	Tax loss	Net cost	NPV
4	195 000	33 222	228 222	
5	195 000	24 917	219 917	
6	440 800	18 680	459 480	
7	440 800	14 016	454 816	
8	440 800	42 048	482 848	
9	440 800		440 800	
10	440 800		440 800	1 547 977
11–15	as before			322 106
16–20	as before			225 974
21	as before			685 912
			NPV	2 781 969

Still showing a balance in favour of renting.

Note that it is the difference in the interest expectations, rather than just tax effects, which result in the expected benefits. It can, in general terms, be expected that company borrowing costs will be higher than property yields, so the example illustrates a normal situation.

Accounting effects of purchase vs lease

There are a number of perceived problems and issues surrounding the property ownership and valuation when presented in company accounts which disguise its true worth/costs and contribute to the lack of shareholder awareness. The London Business School report[4] indicates that the influence of property is considerable.

Example

A Retail chain with substantial property ownership is compared with a similar-sized chain with a leasing policy.

Company A borrows the full cost of a property which cost £5000 at a fixed rate of 20% p.a. Company B leases at a rent equal to 10% of cost of purchase.

Profit and loss

	Company A	Company B
Turnover	10 500	10 500
Cost of sales	6 000	6 000
Gross Trading Profit	4 500	4 500
indirect incl. rent	3 000	3 500
net profit	1 500	1 000
% on borrowing	1 000	
	500	1 000
dividend @ 25% net profit	125	250
Carried Forward	375	750

Balance Sheet

Fixed Assets	6 000	1 000
Current assets less current liabilities	375	750
Net assets	6 375	1 750
long-term creditors	5 000	
	1 375	1 750
Shares	1 000	1 000
From Profit and Loss	375	750
	1 375	1 750

The longer term

In Year 5   (rental and capital growth at 8% p.a;
rent = £735, Capital Value now = £7350)

Gross profit from sales	7 000	7 000
indirect costs	4 000	4 735
net profit	3 000	2 265
% on borrowing	1 000	
	2 000	2 265
dividend (× 25%)	500	565
CF	1 500	1 700

Balance Sheet (year 5)

Fixed assets	2 000	2 000
Property (revalued)	7 350	
	9 350	2 000
Current assets less liabilities	2 500	2 700
	11 850	4 700
Long-term creditors	5 000	
	6 850	4 700
Share capital	1 000	1 000
Capital reserve	2 350	
P&L A/c	3 500	3 700
	6 850	4 700

In another 5 years the cost of B's rent will have overtaken A's borrowing costs. A is a moderately geared company that can increase its capacity for equity by raising the dividend in line with the revaluation. By year 5, Company A is beginning to reap the benefits of the property acquisition because it offers security for a loan *if it needs one*. Company B is ungeared, but must rely on corporate reputation if it needs a loan.

Dividend policy will need to be examined, but in the example Company A's dividends are growing at a faster rate.

Depreciation

If Company A includes a figure for the annual depreciation of the property asset, the position changes again.

Year 5

Net Profit	3 000	2 265
interest on borrowing	1 000	
depreciation @ 4% value	294	
	1 706	2 265
dividend at 25%	426	565

Clearly if the increase on revaluation in Company A can be incorporated as an annual profit, then the dividend can be increased in line with Company B.

Value at the beginning of the year	
(5 000 × 4 years @ 8%)	6 802
Value at end of the year	7 350
Growth for year	550
25% dividend	137.5
Total dividend	564

The nature of most companies prevents the shareholders participating in annual growth of capital value unless they realize the value of their shares. If Company A were to distribute a greater percentage of their profit as a dividend, this would have an impact on their day-to-day cash flow. The evidence is that the value of the company is reduced by holding capital assets because of the market discount to net asset value.

Gearing and ratios

The above 'accounts' have clearly been simplified to extract all other differentials. None the less, they show very clearly the effect on the accounts of a single property decision.

Key ratios which are clearly not comparable are:

	Company A	Company B
Gearing	nearly 50%	ungeared
ROS	10%	5%
Dividend return	$12\frac{1}{2}$ %	25%
Net Equity Growth	× 5	× 2.7
ROCE	11.1%	22.2%

Gearing will be affected by the amount of debt rather than equity. The provision of depreciation will further distort the outturn.

i.e. the companies are virtually incomparable, yet they are both carrying on similar businesses with the same trading profits. In a few years' time,

Company A will increase its capital reserve through revaluation and the borrowing cost will remain fixed; Company B will be paying more rent. Company A will also have further borrowing capacity. Because the nature of the leasing commitment is not disclosed, the shareholders cannot interpret the results.

Leasehold reporting

A further investigation will show the effect on Company B of the lease structure being construed as a finance lease:

Year 1

Gross profit	£4500
Net profit	£1000
Capitalized rentals	£500 × PV of £1 p.a. for 25 years @ 10%
	£500 × 9 = £4500
Net creditors	£4500

Year 5

In year 5 the rental cost will have increased to	£735 (see above)	
Capitalized @ PV of £1 p.a. 20 years @ 10%	8.5	
	= £6 257	

Note the rising total cost. It is only in the last 10 years that the total cost begins to fall.

If the lease is bought at a peppercorn rent, then this will also represent the value of the lease over time.

The above propositions clearly show that even if all other things are equal, the purchase or leasing of property, its financing by debt or equity, its depreciation and revaluation will all quite significantly affect the profile which a company presents to the equities market.

Return on capital employed

Company A has a substantial asset base, but it has the appearance of providing low returns on investment compared to Company B (unless it increases the dividend at the expense of both cover and cash flow). New equity shareholders will therefore be less attracted to it. It does, however, provide a better financial base for debt capital and continued expansion. Indeed, in the longer term, Company B may be forced into buying more fixed assets to provide a security for further capital. Revaluation accentuates this problem where dividends cannot be paid out of the capital growth.

Price earnings ratio

The market would have to consciously analyse all the variables to arrive at a P E ratio which credits Company A with the capital growth, despite the lower dividend (or cover). Indications are that the discount to net asset value shown in company share prices will not adequately compensate for this.

Revaluation

Valuation to market value in existing use will disguise the fact that, for instance, Company A may be occupying property which is underused and has potential for development worth £X. There is no obligation to incorporate this in the accounts.

Particularly the Depreciated Replacement Cost method of Valuation does not represent the price which the asset could be sold for. This may not be a serious consideration for plant and machinery which are known to depreciate rapidly and have little value outside the business, but has serious implications when the asset is growing fast in value.

These issues can lead to takeovers of the company. These will either be to grow an existing business by taking advantage of underused assets, or to asset strip.

Property investment

The nature of property as a quasi-investment is not recognized by the company or shareholders unless the company is restructured to provide separate property financing.

Asset stripping

Historically asset stripping has consisted of buying up a company at share face value which contained an asset (usually property) which was undervalued.

Example: asset stripping

Company S owns factory premises purchased in 1950 for £5 000 (current worth in existing use £200 000). It has been depreciated on a straight-line basis over 50 years:

Balance Sheet	Fixed asset	1 000
	Plant and machinery	50 000
	Net current assets	30 000
		81 000
Financed by	Shares and reserves	55 000
	P&L	26 000
		81 000

P&L income, say, £20 000 p.a. results in current dividends of, say, £5 000 and a market price therefore of about £50 000.

Obviously, the purchaser of all the shares can afford to push up the price to acquire the £200 000 asset.

Even if the property value were incorporated into the accounts, the usual market discount to net asset value would enable a predator to make substantial inroads into a company's shares.

Revaluation is not necessarily the answer.

In the first instance, as a defence to purchasers the directors – if they are able – may take the action of revaluation, increasing the capital valuation of the company to £276 000. The dividends earned may not be adequate to support this. If shareholders discount net asset value, they are unlikely to be tempted to low returns by the promise of capital growth of a property, which may never be realized:

$$5000/276\,000 \times 100\% = 1.8\% \text{ ROCE}$$

A supplementary dividend can be paid, but only at the expense of reserves. A total return of even 5% would look poor for a trading corporate, yet for a property investment company would not be unrealistic.

Consequently, the 'correct' action may well have been overlooked at an earlier stage. This property would earn more for the company if it were sold and leased back, or sold in favour of a less valuable property. Early action would have forestalled the takeover.

Alternative use values

Alternatively, although £200 000 is the current market value of the property in its existing use, this valuation is not necessarily the open market value in an alternative use. A site of a retail centre, for instance, might be worth £2 000 000. Revaluation is no defence against that alternative:

(a) Property is such a major asset of a company that a predator company may purchase to secure markets or outlets for its own use.
(b) Public houses, for instance, which are largely owned by the big brewers were a target. Transactions which have taken place have enabled purchasers to secure further outlet for their own 'beers'. This may or may not affect the trade of the purchased company, but in most cases the smaller company has been swallowed by the larger.

Normally the subject company will fight to avoid takeover for this purpose – as Scottish & Newcastle did to avoid the giant Foster's empire taking over its Northern chain of outlets. (Revaluation was not sufficient to deter Foster's and it was in fact the Monopolies and Mergers Commission[5] which 'saved' this group.)

The accounting conventions in these cases have not assisted sensible planned strategic asset management. Property therefore becomes a pawn to be juggled for financial reasons.

'Off' and 'on' balance sheet

Companies have used two main financial strategies.

1. To take property (among other assets) 'off balance sheet'. This is usually done through creating Joint Venture Companies or by solutions such as sales and leasebacks.

By taking the property off the balance sheet, differences in the financial structure of the two companies, A and B in the examples, are partially removed. The joint venture company will charge Company A with rent, so that its operating accounts will be similar to Company B's. The consolidated accounts will take in the net assets of joint venture, which will disguise the financing arrangements.

Wholly owned subsidiaries now will have to be incorporated into the accounts in the same way as a division of the company, with the new accounting standards looking for substance rather than form:

2. To set up separately accountable property divisions. These are required to act as property traders and to make profits in their own right. Profits from trading (but not from unrealized capital growth) will appear as profit from normal activities and not extraordinary items or investment gains and can therefore be included in the profit for dividend purposes.

17.4 TAXATION EFFECTS OF PROPERTY DECISIONS

Any property sale, or realization of capital value, will potentially give rise to a tax liability. One may rightly criticize property valuation principles, in that the effects of taxation are almost entirely ignored. The derivation of market yields on a comparison basis leads to the use of gross yields. Valuation principles usually assume that roll-over relief will take effect – i.e. no tax will actually be paid. (Accounting for the liability for deferred tax is provided for in SSAP.15.)

Other major taxes which will affect valuation are capital allowances, corporation tax, VAT and the uniform business rate, and such schemes as the Business Expansion Scheme, Urban Development Grant, etc. The taxation differential as applied to property transactions will lead to advantages in some forms of finance deals.

The problem with the lack of explicit definition of taxation aspects in the valuation is the inability of the reader to interpret the difference between those valuations where the tax effects have become implicit in the yield because they affect the whole market and those which are personal to a particular occupier/owner and where additional account needs to be taken in the purchase decision – e.g. corporation tax, grants, reliefs, etc.

The purchaser of a property asset can secure the following:

25% WDA on plant and machinery – reducing balance
4% WDA on industrial premises

hotels　　　　　　　　　⎱
agricultural buildings　⎰　　　　　straight line

100% FYA on industries in enterprise zones
scientific research buildings

Provided the agreement of the freeholder can be obtained, these can also be passed on to a lessee.

As the typical freeholders are non-tax or preferentially taxed investment funds, there is scope for negotiation here. However, occupiers have frequently not sought the advantages of transferred allowances.

A good illustration of the lack of advantage secured to occupiers is shown in an example taken from 1984 when small industrial units earned 100% FYAs.

Example:

In 1984 a small unit available with 100% FYA
Cost of building £60 000 land price £5000
Market rental value £7000
Without allowances, these would have had a value around £56 000 – i.e. the development could not have taken place.
With the FYA, the purchasers were paying around £100 000 for such buildings

(7000 × PV £1 p.a. in perp. @ 7% = £100 000)

The returns were apparently unacceptably low for a property carrying the risk of small industrial units, but accounting for the First-year Allowances were equal to 13.6%[6] (for 52% taxpayers).

Where the purchaser was an owner-occupier, clearly there was advantage. However, many (if not most) purchasers were investors (particularly high tax paying individuals), taking advantage of the very high net of tax returns. The letting rents they achieved did not vary from those achievable without the IBA – indeed, the only factor affecting rental was the increased supply which satisfied demand. Business occupiers were therefore paying full rents and not obtaining any benefits from the taxation advantages. The gainers were land-owners, developers and investors who 'shared' the profits at the expense of businesses. This was clearly not the government intention, and policy was switched to BES rather than 100% FYA schemes.

Currently the major capital allowances advantage lies in the 25% WDA for plant and machinery. Mr Justice Lindley, in *Yarmouth* v. *France*, said: 'There is no definition of plant in the Act but in its ordinary sense it includes whatever apparatus is used by a businessman for carrying on his business . . . all goods and chattels fixed or moveable, live or dead . . .' In particular, it has however been held to include plant for lifts, cabling, airconditioning and the false ceilings, etc. provided for the ducting. This can be expected to account for 30% of the cost of a refurbishment of a new commercial building.

17.4.1 Corporation taxes

These do not have any particularly special property effect to occupiers, except to note the advantage to a company in the year of creating a separate property company, where the unsymmetrical recognition of rental income and expenditure (i.e. receivables vs accruals) enables a company to take permanent cash flow advantage by deferring tax on rent receivable to the next financial year.

Property for investment, however, is bought by both taxpayers and non-taxpayers competing in the market. There are some situations, particularly those where a property opportunity will yield a substantial capital gain over a short time span, where the added value to the taxpayer should be recognized.

17.4.2 Tax relief on interest

Tax relief on interest payments are limited for trading/investment companies. These are the very subsidiaries that are now being set up, so they need to be aware of some of the limitations.

17.4.3 Relief for interest

	Trading company	Investment company
Bank Interest	Trading Expense	Charge on Income
Annual Interest	Charge on Income	Charge on Income
Short Interest	Trading Expense	Not Deductible!

Note that an empty company with nothing to set the interest against will also lose the allowance. The Inland Revenue will usually recognize a property holding company only as an investment company.

17.4.4 Lease Purchases and Premiums

Businesses can be tempted to purchase leases of property – i.e. the right to occupy for a limited period of time at a low or very low rent. The purchase price is, in effect, the capitalized (discounted) profit rental. Except in cases of short leases where the lessor is required to pay Schedule A tax on some part of the lease premium[7] which, in turn, provides a tax-free allowance to the lessee approximately equal to his capital depreciation, there are *no* tax allowances on the price paid for a lease. This therefore is a strong disincentive to purchase as capital outlays cannot be recovered against tax, despite the depreciating nature of the assets.

17.4.5 Capital gains taxes

Capital gains charges are rarely incurred on occupational business property as on sale any charges can normally be rolled over into the business, unless there is clear net disinvestment. However, in exploring strategic asset management, decisions might be made involving disposal of valuable surpluses or transfer of property to a holding company which can on occasion lead to tax payments.

Capital gains taxes are calculated in the usual way (value on sale, 1982 value × indexation). The particular pitfall is in property which has achieved its redevelopment potential since 1982 and which therefore can lead to a high tax charge. Values at 1982 are derived from comparables. It is as well to be aware that the District Valuer keeps records of all property transactions and is therefore exceptionally well-informed.

17.4.6 Value added tax

Property has always been 'protected' from VAT with many transactions zero-rated or exempt.

Current major changes as a result of EC rulings result in a confusing combination of tax-exempt and standard rating (sometimes derived from an option to tax). A landlord who opts to tax a property letting *may* be providing a cash flow advantage to the occupier by enabling him to set VAT on service charges paid – previously exempt – against VAT inputs while, in turn, enabling himself to offset the VAT on the building costs against rental income VAT. The greatest potential for gains for both parties is on office blocks, which have high service charges. Unfortunately, they create a problem where the lettings are to the financial services sector because tenants are unable to take advantage by VAT registration. It is expected that a two-tier market in commercial offices will develop, dependent on the likelihood of subsequent tenants being unable to register.

However, gains to the occupier can also be achieved on business park/light industrial estates and shopping precincts where previously service charges were tax-exempt (i.e. VAT on the services had to be paid by the landlord but could not be reclaimed and were therefore recharged to the tenant in their entirety). The landlords of these estates should now always elect to tax their rental income.

17.4.7 Uniform business rate

The effects of the National Non-domestic Rate taxation have been far-reaching in creating differentials across the country. In turn, this has affected the property market both in the regions and the property types which have been particularly hard-hit.

Phasing relief is allowing new rates to be bought in slowly over five years for existing occupiers, but a particular hazard which may go unnoticed is the loss of phasing relief where a property changes hands, even if the same group of people occupy following a company restructuring. The Uniform Business Rate is one-third to one-half the cost of the market rent and can in some businesses be a large proportion of net profit. This may affect the ability to collect rent.

Example: The taxation effects of a simple *Sale and Leaseback*

This transaction is divided by the Inland Revenue into two separate transactions.
Thus:
– on sale, capital gains tax arises. This may include options, etc., as well as actual cash payments. Any balancing charges deriving from previous capital allowances will need to be accounted for. Stamp duty can be avoided by setting up a separate company.
– on leaseback, rents must not be in excess of market value. If they are, they will be treated as capital and deductions will be restricted. A rent-free period is not satisfactory because subsequent charges cannot be relieved against the rent-free period. Consequently, where less than the full capital value is being raised, the company should seek a long-term reduction in the rent payable.

VAT

– The owner can elect to tax (or VAT will be at standard rate if property is less than 3 years old). Remember the option is irrevocable if in later years the company should diversify into financial services.

Uniform Business Rate (UBR)

– On creation of a new tenancy, the UBR becomes payable in full, without phasing relief.

17.5 FINANCIAL STRATEGIES

The method of holding, as opposed to the decision to use real estate, is held to be a financing solution rather than an investment one, although this can be questioned in the light of various emerging company structures designed to secure the investment advantage from property. Many of the financing mechanisms described are too new to be evaluated in practical, rather than theoretical, terms.

A number of financial alternatives have been tried by corporate management in order to fulfil their property objectives:

Property financing objectives

- To maximize shareholder wealth.
- To minimize cost of funding.
- To obtain control/flexibility over use of premises
  (i.e. ability to expand/contract space, ability to alter use of space, ability to upgrade accommodation, ability to prevent control passing into 'unfriendly' hands and ease of administration/management).
- To take advantage of real estate growth.
- To maximize tax allowances.
- To have the lowest balance sheet impact.
- To provide protection from asset strippers.
- To match funding in terms of earnings.
- To liquidify non-earning assets.
- To extend tenor of finance.
- To solve the 'reverse yield gap'.
- To remove the distorting factor of property growth from the corporate accounts of the core business.

It has been suggested[9] that in the USA there are over 500 types of financial instrument available.

Financing and Holding Arrangements and Strategies:

Structures:

On Balance Sheet – Property Holding Companies

Off Balance Sheet – Joint Ventures; Companies and Partnerships; (Non-consolidated subsidiaries, controlled non-subsidiaries and options are now the subject of legislation and Accounting Standards and are unlikely to be achieved off balance sheet in the future.)

Financing:

On Balance Sheet – Equity, Debt, some Sales and Leasebacks.

Off Balance Sheet – Sale and Leaseback, Joint Ventures, Non-recourse loans.

Many of the most innovative strategies and arrangements have been pioneered by cash-hungry property companies, with APUTS (Authorized Property Unit Trusts), FRNs (Floating-rate Notes), defer and accrue loans, securitization and an amalgam of discounted debt and PUT and CALL options.

The number of possible permutations and combinations of the various arrangements and strategies are so great that only one, well-established 'off balance sheet structure' can be explored here.

Off balance sheet structures: sale and leaseback

This is an old-established technique used by the occupier of a property which enables the occupier to raise funds and perhaps make profit on sale, whilst retaining use. The cost of renting is likely to be lower than the borrowing cost of the cash raised and it is therefore a cheap source of capital in the short term. Its cost should be measured against the cost of alternatives.

Assuming the sale and leaseback are at fair market value, accounting is simple:

- it is dictated by SSAP 21 which distinguishes between a finance lease and an operating lease.
  A finance lease 'transfers substantially all the risks and rewards of ownership of an asset to the lessee'. All finance leases have to be capitalized by the lessee. An operating lease is any other lease, and does not have to be capitalized by the lessee. (If the objective is partially to take the property off the balance sheet, operating lease status will have to be sought).
- the property disappears from retailer's balance sheet.
- net sales proceeds are included in the balance sheet as cash or debtors.
- the profit (or loss) is recognized as the difference between net sales proceeds and book value.
- rents payable are charged as overheads over the term of the lease.

Accounting effects of Sale and Leaseback

	Before	*After* Finance Lease A	*After* Operating Lease B
Fixed Asset	100m.	100	80m.
Current assets			
less liabilities	5	25	25
	105	125	105
Loans	40	60	40
Equity	65	65	65
Gearing	40/105	60/125	40/105

Gearing remains consistent following a sale and leaseback if it is classified as operating, whereas security for further borrowing is reduced in both cases. There is no apparent liability to repay the loan, but profits will be reduced by the future rental payments, which has longer-term implications for future expansion. (Recent refinements include 'put' and 'call' options, so the seller can preserve his interest in the capital growth and the lender can be protected from capital losses.)

Additionally, the leasing owner is disadvantaged by not being able to take advantage of a tenants' market (as tenants are currently able) as he is both the freeholder and lessee. He will be seeking the best possible purchase price, but his subsequent position as a tenant is not one from which he can strongly argue terms; a higher purchase price must only be argued on the basis of a lower yield, not a higher rental value, or the rent payable will rise.

Example: evaluation of sale and leaseback

Company W wish to raise £20 million for expansion but have had to rule out further borrowing at current interest rates.

The money could be raised by an equity rights issue which at current WACC is 15% net or by selling a number of retail outlets for £20 million and leasing back for 125 years at an initial rent of £1 000 000 subject to 5-yearly rent reviews.

Assuming that average annual growth in rents and property values is likely to be 8% p.a., that dividends will rise by the same amount and that the company will increase in value by 10% p.a. following the new investment.

Year	Outflow (Rent)	Inflow	Net (tax)	PV @15%	NPV
0		20m.	20m.		20m.
1–5	£1m.		(650 000)	3.3	(2 178 000)
6–10	£1.46m.		(949 000)	1.65	(1 566 000)
11–15	£2.145		(1 394 000)	0.82	(1 759 000)
15	Value of asset forgone*		(63 443 000)	0.14	(8 966 000)
			Total cost	14 470 000	
			NPV	5 530 000	

* Current Cap Value × amount of £1 15 years @8%.

This makes the sale and leaseback an attractive option, provided the investment of the £20 million is generating profit to sustain the existing dividend, despite the rental payments. The sale and leaseback is *not* a short-term working capital injection for a business in trouble – the purpose for which it is often used.

Apart from the other clear financial costs – i.e. costs of negotiating the deal, taxation, etc. – which can easily be incorporated, there is the significant longer-term effect of the decision on the firm's balance sheet, its credit rating for future loans and the cash flow position of the firm.

Variations on a theme: sale and leaseback schemes

A Top/Bottom slice arrangement exists where a property is sold for less than full market value, on condition that rent paid is less than rack rental.

The occupier therefore retains a proportion of the equity depending on the size of the slice sold, i.e.:

Property worth F/H £1.5m.
Occupier wishes to raise £1m.

Sale price £1m. F/H
Leaseback 125 years @ 66% of rack rent (reviewable every 5 years
to 66% of current rack rent).
A *lease and leaseback* is similar to the above.
Property could be sold on a long (125 year) lease for £1 million and leased
back at 66% rack rental.
Note that in these methods both the rent paid and the profit rental increase by
the same percentage. There is no gearing effect.

Geared top slice arrangements

The 'rent' payable can be linked to the amount of finance raised with the rental
growth 'geared' between the parties:

Example: Geared top slice

Amount raised	£1m.
Initial rent @ 6.5%	65 000
Rental value =	97 000
% of rack rent =	66.6%
At first review rental value	150 000
rent payable	65 000
+ 50% of growth in rental growth	52 000 ÷ 2
	26 000
Total rent payable	91 250

The relative advantages of this depend on the growth percentages agreed and
might be associated with a relatively high initial rent.

Sale and finance leaseback

This is a useful method of financing property for the owner's own use.
e.g. The owner grants a bank a 25-year lease for no premium and at
   peppercorn rent.
  The bank develops the building and then leases the building back to the
  owner for 25 years for no premium.
  The lease rentals are calculated to repay bank development costs plus
  interest.
  Clearly, the lease obligations would have to be capitalized by lessee as a
  finance lease.
This structure is usually motivated by tax considerations (the bank shares the
benefit of capital allowances with the owner) but lessees should note that the
finance costs cannot be deducted as rent from annual profits in as much as they
are greater than a rental payment. Where higher payments are demanded, the
excessive rent is rolled forward until such time as rising rental values enable
the whole costs to be used up.

Sale and repurchase

In this variation, two agreements are entered into at the same time – namely, sale, and repurchase:

- The price payable for repurchase is the sale price, plus interest equivalent.
- The intended effect is that seller's balance sheet should include cash received (sale proceeds)
  but not the property or the obligation to repay, and it is the writer's opinion that it is doubtful whether this would be accepted by the company's auditors.
- A further refinement is used by some house builders for show houses: the builder sells some completed show houses to a bank, which grants the builder the right to use the houses for a fee which equates to interest on the funds employed by the bank. The builder commits to sell the houses to the public as agent for the bank and the builder retains all sales proceeds, less sums due to the bank, as its agency fee. (The builder may, or may not, guarantee shortfalls between sales proceeds and the bank's commitment.)

Leveraged sale and leaseback

These are not developed in the UK. They involve a fixed 'rental' based on finance rates for a number (usually 10) of years and a reversion to a higher interest rate for the remainder of a term. The British financial institutions prefer the return to be based on the equity of the property market – rental growth has historically led to higher rewards. The British taxation system also disadvantages the finance-based lease.

REFERENCES

1. Gordon, M.J. (1974) A general solution to the lease or buy decision, *Journal of Finance*, **29**, 245–50.
2. Brearley, R. and Myers, S. (1988) *Principles of Corporate Finance*, McGraw-Hill, New York and London.
3. Johnson, R.W. and Llewellyn, W.G. (1972) 'Analysis of the lease or buy decision', *Journal of Finance*, September, 815–23.
4. Currie, and Scott, *The Place of Commercial Property in the UK Economy*, London Business School, London, 1990.
5. Monopolies and Mergers Commission, *The Supply of Beer*, Report, 1989.
6. Colborne, A. (1985) Tax explicit valuations, industrial building allowances, *Journal of Valuation*, **2**(2), 117–24.
7. Colborne, A. (1978) 'Tax explicit valuations, lease premiums', *Estates Gazette*, September, 247EG 1065.
8. Chartres, M. (1990) Off balance sheet finance, *Journal of Property Finance*, **1**(1).

Bibliography

Advisory Council for Applied Research and Development (1980) *Information Technology*, Her Majesty's Stationery Office, London.

Bank Administration Institute (1968) *Measuring the Investment Performance of Pension Funds for the Purpose of Inter-fund Comparisons*, BAI, Park Ridge, Illinois, USA.

Barber, C. (1988a) Performance evaluation, *Estates Gazette*, 27 August, pp. 53–4.

Barber, C. (1988b) Unravelling the indices mysteries, *Chartered Surveyor Weekly*, 1 September, p. 32.

Baum, A. (1987) Risk-explicit appraisal: a sliced income approach, *Journal of Valuation*, **2**, 250–67.

Bon, R., Joroff, M. and Veale, P. (1987) *Real property portfolio management*, discussion paper and symposium summary, The Laboratory of Architecture and Planning, MIT.

Brearly, R. and Myers, S. (1988) *Principles of Corporate Finance*, McGraw-Hill, New York and London.

Brew, J. (1968) The measurement of portfolio performance – internal assessment, private seminar paper, SIA, July.

Buchan, J. and Koerigsberg, E. (1963) *Scientific Inventory Management*, Prentice-Hall, Englewood Cliffs, NJ.

Burch, J.G., Jr, Starter, F.R. and Gradmitski, G. (1979) *Information Systems: Theory and practice*, 2nd ed, John Wiley and Sons, New York.

Byrne, P. and Cadman, D. (1984) *Risk, Uncertainty and Decision-making in Property Development*, E. and F.N. Spon, London.

Cabinet Office Information Technology Advisory Panel (1983) *Making a business of information*, HMSO, London.

Carter, E.E. (1985) Change and technology in society, in *Introduction to computer assisted valuation*, (ed A. Woolery and S. Shed), Oelgeschlager, Gunn and Hain.

Chapman, H., Wyatt, A. and Thompson, J. (1980) Measuring property performance, *Chartered Surveyor*, **112**, 444–6.

Chartres, M. (1990) Off balance sheet finance, *Journal of Finance*, Autumn, **1**(1).

Cocks, G. (1972) An objective approach to the analysis of portfolio performance, *Investment Analyst*, **34**, 3–7.

Colborne, A. (1979) Tax explicit valuations, lease premiums, *Estates Gazette*, September, 247EG 1065.

Colborne, A. (1985) Tax explicit valuations, industrial building allowances, *Journal of Valuations*, **2**(2), 117–24.

Computer Briefing (1988) Risk simulation software – predict! *Journal of Valuation*, **2**(6), 313–17.

Conglong, G. (1972) The FT Actuaries All Share Index – a commentary, *Investment Analyst*, December.

Coyle, R.R. (1972) *Decision Analysis*, Nelson, London.

Crosby, N. (1988) An analysis of property market indices with emphasis on shop rent change, *Land Development Studies*, March.

Currie, D.A. and Scott, P. (1990) *The Place of Commercial Property in the UK Economy*, London Business School.

Darlow, C. (ed.) (1988) *Valuation and investment appraisal*, Estates Gazette, London.

Dietz, P.O. (1966) *Measuring Investment Performance*, Columbia University Press, New York.

Dilmore, G. (1981) *Quantitative Techniques in Real Estate Counselling*, Lexington Books, Lexington, Mass.

Duckworth, W.E. *et al.*, (1977) *A Guide to Operational Research*, Chapman and Hall, London.

Eade, C. (1992) IPD index lays the groundwork for raising property's profile, *Chartered Surveyor Weekly*, **38**(3), 24–5.

Eadie, D. (1973) A practical approach to the measurement and analysis of investment performance, *Investment Analyst*, **37**, December, 12–18.

Edelman, D.B. (1986) *Statistics for property people*, Estates Gazette, London.

England, J.R., *et al.* (ed), *Information systems for policy planning in local government*, Longman London.

European Federation of Financial Analysts Societies (1965) *Closing Report of the Third Congress*, EFFAS, London.

Finlay P. and Tyler S. (1990) *Performance Measurement of Property Investments*, Aoteaora Press, Nottingham.

Firth, M. (1975) *Investment Analysis*, Harper and Row, New York and London.

Fraser, W.D. (1984) *Principles of Property Investment and Pricing*, Macmillan, London.

Frazer, W. (1985) The risk of property to the institutional investor, *Journal of Valuation*, **1**(4), 45–59.

Frost, A.J. and Hager, D.P. (1986) *A General Introduction to Institutional Investment*, Heinemann, London.

Gibson, E.J. (ed.) (1979) *Developments in Building Maintenance*, Applied Science Publishers, London.

Gilliand, A.B. (1962) Measuring ordinary share portfolio performance, *Investment Analyst*, August, **3**, 30–5.

Gordon, M.J. (1974) A general solution to the lease or buy decision, *Journal of Finance*, March, **29**, 245–50.

Gronow, S. and Scott, I. (1986) Expert Systems, *Rating and Valuations*, April, 111–22.

Hager, D.P. and Lord, D.J. (1985) *The Property Market, Property Valuations and Performance Measurement*, Institute of Actuaries, London.

Hall, P.O. (1981) Alternative approach to performance measurement, *Estates Gazette*, 19 September, p. 935–8.

Hall, P. and Hargitay, S. (1984) Property portfolio performance – a selected approach, *Property Management*, **2**, 218–29.

Hargitay, S.E. (1985) *Property Portfolio Analysis Package*, Private Publication, Bristol.

Healey and Baker (1986) *Property Rent Indices and Market Editorial (PRIME)*, Healey and Baker Research, London, June.

Hertz, D.B. (1964) Risk analysis in capital investment, *Harward Business Review*, January–February, 95–106.

Hetherington, J. (1980) Money and time weighted rates of return, *Estates Gazette*, **256**, 1164–5.

Hillier, F. (1963) The derivation of probabilistic information for the evaluation of risky investments, *Management Science*, April, **9**, 443–57.

Hillier Parker (1979a) *Investors Chronicle/Hillier Parker Rent Index*, Investors Chronicle/Hillier Parker, London, May.

Hillier Parker (1979b) *The Relationship Between Shop Rents and Town Centre Size*, Research Report No. 3, Hillier Parker Research. London.

Hillier Parker (1983) *Portfolio Analysis*, Hillier Parker Research, London.

Hillier Parker (1985) *Secondary Rent Index*, Research Report No. 7, Hillier Parker Research. London.

Hillier Parker (1987) *Investors Chronicle/Hillier Parker Rent Index*, Investors Chronicle/Hillier Parker, London, November.

Hillier Parker May and Rowden (1983) Portfolio Analysis, Hillier Parker May and Rowden Research, London, August.

Hsia, M. and Byrne, P. (1988) *Automated property performance analysis: considerations on the development of a PC based information system prototype*, Working Papers in Land Management and Development, University of Reading.

Hull, J.C. (1980) The evaluation of risk in business investment, Pergamon, Oxford.

Humphries, B. (1984) The crisis in land management, in *The Decision Maker and Land Information Systems* (eds A.C. Hamilton and J.D. McLaughlin), papers and proceedings form the FIG International Symposium Edmonton, Alberta, Oct., p. 10.

Hymans C. and Mulligan, J. (1980) *The Measurement of Portfolio Performance*, Kluwer, London.

IPD (1992) *The IPD Annual Review, 1992*, Investment Property Databank, London.

Jensen, M. (1969) Risk, the pricing of capital assets and the evaluation of investment portfolio, *Journal of Business*, XLII, April, 167–247.

JLW (1984) *JLW Index Explanatory Notes*, Jones Lang and Wootton, London.

JLW (1987) *JLW Property Index*, Jones Lang and Wootton, London, Spring.

Johnson, R.W. and Llewellyn, N.G. (1972) Analysis of the lease or buy decision, *Journal of Finance*, September, 815–23.

Jones Lang and Wootton (1980) *Property Performance Analysis System*, Jones Lang and Wootton, London, January.

Jones Lang and Wootton (1982) *Property Investment Performance over 20 years*, Occasional Paper, Jones Lang and Wootton, London, Summer.

Keen, P.G.W. and Scott Morton, M.J. (1978) *Decision support systems: An organisation perspective*, Addison-Wesley, New York.

Kerridge, D.S. (1987) *Investment – a Practical Approach*, Pitman, London.

Kirkwood, J. (1985) Information technology: its impact on real estate valuation and management, in *Introduction to Computer Assisted Valuation* (eds A. Woolery and S. Shea) Oelgeschlager, Gunn and Hain, p. 2.

Kohlhepp, D.B. (1982) Computers in appraising: applications, problems, and possible solution, *The Real Estate Appraiser and Analyst*, 22–5.

Levy, H. and Sarnat, M. (1982) *Capital Investment and Financial Decisions*, Prentice-Hall, Englewood Cliffs, NJ.

Lumby, S. (1981) *Investment Appraisal and Related Decisions*, Nelson, London.

Makower, M.S. and Williamson, E. (1975) *Operational Research*, Hodder and Stoughton, London.

Malca, E. (1973) *Bank Administered Co-mingled Pension Funds*, Lexington Books, Lexington, Mass.

Mandell, S.L. (1985) *Computers and Data Processing*, West Publishing, St Paul, USA.

Markowitz, H. (1959) *Portfolio Selection: Efficient Diversification of Investments*, Wiley, New York and London.

Marshall J.B. (1980) Pension fund performance – a new approach, *Investment Analyst*, April, **56**, 14–20.

Mason, R. (1975) Portfolio management, *Estates Gazette*, **236**, 663–8.

Mason, R. (1980) Performance measurement, *Estates Gazette*, 256, 1091–5.

Merrett, A.J. and Sykes, A. (1974) *The Finance and Analysis of Capital Projects*, 2nd edn, Longman, London.

Messner, D. and Chapman Findlay, M. (1975) Real estate investment analysis: IRR versus FMRR, *Real Estate Appraiser*, July–August, **40**(4).

MGL/CIG (1987) *The MGL–CIG Property Index 1978–1986*, Morgan Grenfell Laurie/Corporate Intelligence Group, London.

Morrell, E. D. (1991) Property performance analysis and performance indices: a review, *Journal of Property Research*, no 8, 29–57.

NAPF (1988) *NAPF Investment Committee Annual Report*, National Association of Pension Funds, London.

NAB Research Project on *Essential data requirements for efficient property asset management*, Department of Surveying, Bristol Polytechnic.

Newell, M. (1985) The rate of return as a measure of performance, *Journal of Valuation*, **4**, Autumn, 130–42.

Nijkamp, P. and Rietveld, P. (eds) (1986) *Information systems for intergrated regional planning*, North Holland, Amsterdam.

Porter, M.E. and Millar, V.E. (1985) How information gives you competitive advantage, *Harvard Business Review*, **63**(4), 149–60.

Report of a committee chaired by Lord Chorley (1987) *Handling Geographic Information*, Her Majesty's Stationery Office, London.

Richard Ellis (1983) *Property Market Indicators*, Richard Ellis, London.

Richard Ellis/Wood McKenzie & Co. (1980) *Introduction to the Property Performance Service*, Richard Ellis/Wood McKenzie, London.

Robicheck, A. (1975) Interpreting the results of risk analysis, *Journal of Finance*, December, **XXX**(5), 1384–6.

Robinson, J. (1987) Cash flows and risk analysis, *Journal of Valuation*, 2(5), 268–89.

Rowe and Pitman, (1982) Property indices: do they make sense?, *Journal of Valuation*, **1**(2), 197–201.

Sharpe, W.F. (1970) *Portfolio Theory and Capital Markets*, McGraw-Hill, New York and London.

Sharpe, W.F. (1985) *Investments*, 3rd edn, Prentice-Hall, Englewood Cliffs, NJ.

Sherman, B. (1985) *The new revolution the impact of computers on society*, John Wiley and Sons, New York.

Society of Investment Analysts (1974) *The Measurement of Portfolio Performance for Pension Funds*, SIA, London.

Sprecher, C.R. (1978) *Essentials of Investments*, Houghton Mifflin, New York.

St Quintin (1982) *Compas Computerised Property Appraisal System*, St Quintin Press, London.

Stapleton, T.B. (1986) *Estate Management Practice*, Estates Gazette, London.

The Stock Exchange (1986) *An Introduction to the Stock Market*, The Stock Exchange, London.

Sykes, S.G. (1983) The assessment of property investment risk, *Journal of Valuation*, **1**(3), 253–67.

Taffler, R. (1979) *Using Operational Research*, Prentice-Hall, Englewood Cliffs, NJ.

Thierauf, R.J. and Grosse, R.E. (1970) *Operational Research*, Wiley, New York and London.

Treynor, J.L. (1965) How to rate management of investment funds, *Harvard Business Review*, **43**(1), January–February, 63–76.

Warren, R. (1983) *How to Understand and Use Company Accounts*, Business Books, London.

Wilkes, F.M. (1980) *Elements of Operational Research*, McGraw-Hill, New York and London.

Wiltshaw, D.G. (1987) Expert systems and land development expertise, *Land Development Studies*, **4**, 55–67.

Woodward, J.F. (1975) *Quantitative Methods in Construction Management and Design*, Macmillan, London.

Yeomans, K.A. (1986) *Statistics for Social Scientist* (2 vols), Penguin, Harmondsworth.

Young, M., (1977) Evaluating the risk of investment real estate, *Real Estate Appraiser*, **43**, September–October.

Appendices

Appendix A: Areas in tail of the normal distribution

The function tabulated gives the probability that a standardised Normal variable selected at random will be smaller than a critical value.

$$d = \frac{x_c - \bar{x}}{\sigma}$$

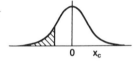

$\frac{(x_t - \bar{x})}{\sigma}$	.00	.01	.02	.03	.04	.05	.06	.07	.08	.09
0.0	.5000	.4960	.4920	.4880	.4840	.4801	.4761	.4721	.4681	.4641
0.1	.4602	.4562	.4552	.4483	.4443	.4404	.4364	.4325	.4286	.4247
0.2	.4207	.4168	.4129	.4090	.4052	.4013	.3974	.3936	.3897	.3859
0.3	.3821	.3873	.3745	.3707	.3669	.3632	.3594	.3557	.3520	.3483
0.4	.3446	.3409	.3372	.3336	.3300	.3264	.3228	.3192	.3156	.3121
0.5	.3085	.3050	.3015	.2981	.2946	.2912	.2877	.2843	.2810	.2776
0.6	.2743	.2709	.2676	.2643	.2611	.2578	.2546	.2514	.2483	.2451
0.7	.2420	.2389	.2358	.2327	.2296	.2266	.2236	.2206	.2177	.2148
0.8	.2119	.2090	.2061	.2033	.2005	.1977	.1949	.1922	.1894	.1867
0.9	.1841	.1814	.1788	.1762	.1736	.1711	.1685	.1660	.1635	.1611
1.0	.1587	.1563	.1539	.1515	.1492	.1469	.1446	.1423	.1401	.1379
1.1	.1357	.1335	.1314	.1292	.1271	.1251	.1230	.1210	.1190	.1170
1.2	.1151	.1131	.1112	.1093	.1075	.1056	.1038	.1020	.1003	.0985
1.3	.0968	.0951	.0934	.0918	.0901	.0885	.0869	.0853	.0868	.0823
1.4	.0808	.0793	.0778	.0764	.0749	.0735	.0721	.0708	.0694	.0681
1.5	.0668	.0655	.0643	.0630	.0618	.0606	.0594	.0582	.0571	.0559
1.6	.0548	.0537	.0526	.0516	.0505	.0495	.0485	.0475	.0465	.0455
1.7	.0446	.0436	.0427	.0418	.0409	.0401	.0392	.0384	.0375	.0367
1.8	.0359	.0351	.0344	.0336	.0329	.0322	.0314	.0307	.0301	.0294
1.9	.0287	.0281	.0274	.0268	.0262	.0256	.0250	.0244	.0239	.0233
2.0	.02275	.02222	.02169	.02118	.02068	.02018	.01970	.01923	.01876	.01831
2.1	.01786	.01743	.01700	.01659	.01618	.01578	.01539	.01500	.01463	.01426
2.2	.01390	.01355	.01321	.01287	.01255	.01222	.01191	.01160	.01130	.01101
2.3	.01072	.01044	.01017	.00990	.00964	.00939	.00914	.00889	.00866	.00842
2.4	.00820	.00798	.00776	.00755	.00734	.00714	.00695	.00676	.00657	.00639

$\dfrac{(x_t - \bar{x})}{\sigma}$	.00	.01	.02	.03	.04	.05	.06	.07	.08	.09
2.5	.00621	.00604	.00587	.00570	.00554	.00539	.00523	.00508	.00494	.00480
2.6	.00466	.00453	.00440	.00427	.00415	.00402	.00391	.00379	.00368	.00357
2.7	.00347	.00336	.00326	.00317	.00307	.00298	.00289	.00280	.00272	.00264
2.8	.00256	.00248	.00240	.00233	.00226	.00219	.00212	.00205	.00199	.00193
2.9	.00187	.00181	.00175	.00169	.00164	.00159	.00154	.00149	.00144	.00139
3.0	.00135									
3.1	.00097									
3.2	.00069									
3.3	.00048									
3.4	.00034									
3.5	.00023									
3.6	.00016									
3.7	.00011									
3.8	.00007									
3.9	.00005									
4.0	.00003									

Appendix B: Extracts from the IPD Annual Review 1992

This year for the first time, the IPD Annual Review includes an analysis of the differences between market and investment returns to property.

Market returns exclude the effects of active management and development exposure, and track the performance of the bulk of the standing investment portfolio in the IPD. In 1991 development spending knocked a full 1.5 points and active management a further 1 point off the − 2.7% market return, dragging investment returns for the year down to − 4.9%

IPD TOTAL RETURNS 1981–1991
The impacts of active management

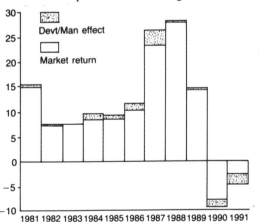

After converging in 1990, sector results began to diverge:

- industrials leading with an 8% return
- shops following at 3.2% for the year
- offices suffering another 10% hit

Less than £1 billion was committed to new investments in 1991, the lowest total since 1985. This was only just over half as much as the proceeds from sales.

If you would like extra copies of the Annual Review, or further details of IPD's range of property information services, please ring Susan James on 071-482-5149

Investment Property Databank, 7/8 Greenland Place, London NW1 0AP
Telephone 071 482 5149
Fax 071 267 0208

IPD market and sector performance
Total return

All properties

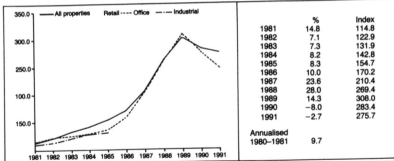

	%	Index
1981	14.8	114.8
1982	7.1	122.9
1983	7.3	131.9
1984	8.2	142.8
1985	8.3	154.7
1986	10.0	170.2
1987	23.6	210.4
1988	28.0	269.4
1989	14.3	308.0
1990	−8.0	283.4
1991	−2.7	275.7
Annualised 1980–1981	9.7	

Retail

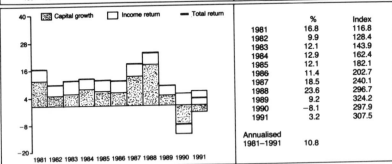

	%	Index
1981	16.8	116.8
1982	9.9	128.4
1983	12.1	143.9
1984	12.9	162.4
1985	12.1	182.1
1986	11.4	202.7
1987	18.5	240.1
1988	23.6	296.7
1989	9.2	324.2
1990	−8.1	297.9
1991	3.2	307.5
Annualised 1981–1991	10.8	

Office

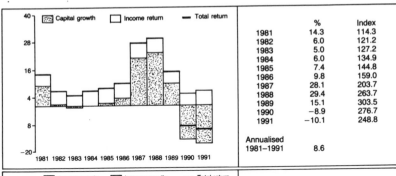

	%	Index
1981	14.3	114.3
1982	6.0	121.2
1983	5.0	127.2
1984	6.0	134.9
1985	7.4	144.8
1986	9.8	159.0
1987	28.1	203.7
1988	29.4	263.7
1989	15.1	303.5
1990	−8.9	276.7
1991	−10.1	248.8
Annualised 1981–1991	8.6	

Industry

	%	Index
1981	12.3	112.3
1982	5.4	118.4
1983	6.3	125.8
1984	6.7	134.2
1985	3.0	138.2
1986	7.2	148.1
1987	22.4	181.3
1988	37.0	248.3
1989	27.4	316.3
1990	−4.5	302.1
1991	8.0	326.3
Annualised 1981–1991	11.4	

Yields

Yield charts

All properties

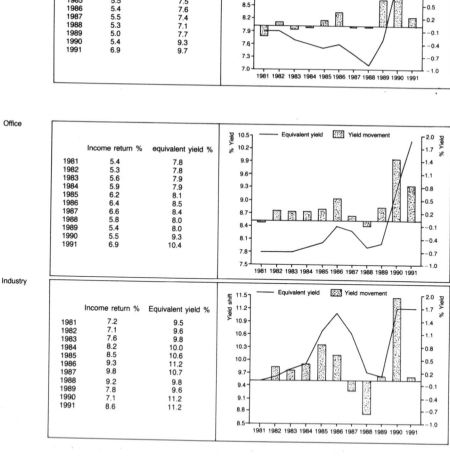

	Income return %	Equivalent yield %
1981	5.6	8.1
1982	5.5	8.1
1983	5.8	8.1
1984	6.1	8.1
1985	6.3	8.2
1986	6.4	8.5
1987	6.6	8.3
1988	6.0	7.8
1989	5.5	8.1
1990	5.7	9.6
1991	7.1	10.2

Retail

	Income return %	Equivalent yield %
1981	4.9	7.9
1982	5.0	7.9
1983	5.3	7.7
1984	5.4	7.6
1985	5.5	7.5
1986	5.4	7.6
1987	5.5	7.4
1988	5.3	7.1
1989	5.0	7.7
1990	5.4	9.3
1991	6.9	9.7

Office

	Income return %	equivalent yield %
1981	5.4	7.8
1982	5.3	7.8
1983	5.6	7.9
1984	5.9	7.9
1985	6.2	8.1
1986	6.4	8.5
1987	6.6	8.4
1988	5.8	8.0
1989	5.4	8.0
1990	5.5	9.3
1991	6.9	10.4

Industry

	Income return %	Equivalent yield %
1981	7.2	9.5
1982	7.1	9.6
1983	7.6	9.8
1984	8.2	10.0
1985	8.5	10.6
1986	9.3	11.2
1987	9.8	10.7
1988	9.2	9.8
1989	7.8	9.6
1990	7.1	11.2
1991	8.6	11.2

IPD PERFORMANCE OVERVIEW AT DECEMBER 1991

ALL PROPERTIES

The 1992 IPD Annual Review is published within days of a cliff-hanging election whose results will probably do little to resolve the uncertainties which currently bedevil the investment property market. The demand economy remains in recession and the overhang of surplus floorspace casts a long shadow upon the route to recovery.

Because of the uncertainty of the current market position, this year's Review has been designed to cover the full spectrum of market returns. At one end, the full impact of a perhaps unwilling participation in large scale development projects and other defensive expenditure pulled the 1991 total investment returns down to −4.9%. At the other, returns to the bulk of the fund's standing investment properties, stripping out the effects of active management to reveal the underlying patterns of market movement, were more than 2 points better, at −2.7%. Thus in 1991, as in 1990, both active (or defensive) management and unavoidable development expenditure each took a point or more off market returns.

The first signs of a turning point are, however, offered by the Monthly Index which has in the past proved a reliable leading edge indicator. This is currently showing a positive 12 month market return. With a low exposure to the City office market and heavy weight in industrials, it has maintained the 2 point lead over the Annual Index which first emerged as early as December 1988.

RETAIL

In 1990 returns to each of the sectors converged upon a 10 year annual rate of just over 11%. 1991 saw the start of a new pattern of diverging performance, but with the hierarchy reshuffled. In 1981 retails took the lead. In 1991 they lagged well behind industrial property, returning only just over 3% for the year and dipping perceptibly into negative rates of rental value growth.

Despite this somewhat lacklustre performance, allocations to retail property increased for the 3rd successive year, moving its weight in the portfolio up to its highest recorded level of close to 40%. This represents a 35% increase in the retail share as compared with the position in December 1981.

Within the retail sector, 1991 saw a bigger range of overall performance scores than ever before, with 'fringe' retails (restaurants, pubs, showrooms etc) producing negative returns and retail warehouses well into double figures and outperforming all other retail categories by a significant margin.

The signs of a reawakening of investment interest in the sector were reflected in yield movements − mostly still outward, but far less so than in 1990 and far less than the adverse shift applied to office property in 1991.

OFFICE

The office sector in fact produced the worst results of all the major property types in 1991. Yields were pushed out by yet another point.

Rental values were cut by 15% in a year in which they were only very modestly trimmed in each of the other sectors. And both capital growth and total return accelerated their rates of decline, whilst these rates were being halted or reversed for retail and industrial units.

The negative impacts of development exposure and active management were also most painfully felt in the office sector. Each took a full $1\frac{1}{2}$ points off the market return figure of −10.1%, dragging total investment returns down to −12.5%. It is clear from a closer scrutiny of the figures that much of the 1991 active management took the form of development expenditure continuing beyond the formal completion of the project.

In regional terms, office returns exhibited very similar patterns to those of 1990, with performance improving noticeably with each main line railway station out of Central London. In the Northern and Scottish regions, returns were typically 2% or better, and they were positive in most centres outside the South East.

INDUSTRY

The industrial sector outperformed all other in 1991, producing returns of 8% for the year and suffering negligible cuts in either capital or rental valuations. After the severe 2 point outward movement in yields in 1990, they were held at just above the 11% mark throughout last year.

Thus 1991 marked the fourth successive year in which industrial property ended ahead of both of the other main sectors. Not since 1987, the year of the Central London office boom, have industrial units been beaten into second place. The broader regional spread of modern warehousing and light industrial stock, coupled with the absence of huge supply surpluses, have meant that returns have been protected against the worst effects of the recession. It may be, however, that if recovery is led by consumption pressure, the lagged response of the manufacturing sector will mean that industrial returns will fade later this year and next relative to those of well located shops.

As with the office sector, industrial performance in the Midlands and the North was well in excess of that of the South East in 1991.

Extracts from the IPD fund comparative report

THE IPD FUND COMPARATIVE REPORT

This Comparative Report sets the performance of the portfolio against that of other property portfolios contributing to the Investment Property Databank.

The Benchmark

In total the Databank now covers 218 funds with total property holdings of £38.5 billion at December 1991. The properties in the databank which are owned by the insurance and pensions funds are estimated to represent 64% of all UK property held by the financial institutions. The majority of funds are valued in December, but a few have March or rolling valuation dates. These are excluded from the IPD benchmarks given in this report, which are based on a total of 157 funds with a December 1991 value of £31.4 billion.

Contents of the Report

The main body of the report is a detailed examination of the fund's characteristics and performance compared with the property market average represented by the IPD benchmark, and the performance of the main alternative UK investments in equities, gilts and property shares.

This is intended to serve three main purposes:

- To give an authoritative benchmark of the relative performance of the fund against that achieved by other property owners.
- To identify the main areas of strength and weakness in the performance of the portfolio, and the influence of the fund's asset mix, trading decisions and development activity.
- To provide the information, ranging from broad sector performance down to returns for each property, which is required for further diagnostic analysis and fund management decisions.

The full comparative charts and tables are preceeded by an Executive Summary, which comments on the main points of interest in the report, highlighting areas where the fund's structure, investment activity or performance are out of line with general trends in the market.

POSITION AMONGST IPD RANKED FUNDS–TOTAL RETURN

% pa Including Transactions and Developments

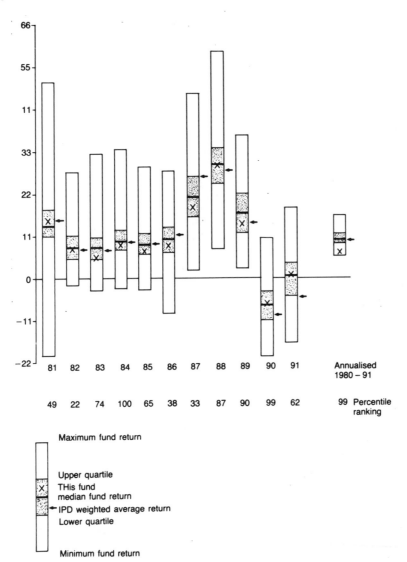

TOTAL RETURN

All Properties, Including Transactions and Developments

Year End 31st Dec	% p.a.		Index Values		Real Returns		IPD Median Fund
	Fund	IPD[1]	Fund	IPD[2]	Fund	IPD	
1980	–	–	100.0	100.0	–	–	–
1981	15.2	15.3	115.2	115.3	2.8	3.0	13.8
1982	7.7	7.5	124.0	123.9	2.2	2.0	8.1
1983	5.4	7.4	130.8	133.2	0.1	2.0	8.1
1984	8.8	9.6	142.3	145.9	4.0	4.8	9.7
1985	7.1	9.2	152.3	159.3	1.3	3.3	8.9
1986	8.6	11.4	165.5	177.4	4.8	7.4	10.2
1987	18.5	26.6	196.2	224.7	14.3	22.1	21.2
1988	29.5	28.2	254.0	288.1	21.2	20.0	29.8
1989	14.3	14.7	290.3	330.4	6.1	6.5	17.1
1990	– 6.6	– 9.6	271.1	298.8	– 14.6	– 17.3	– 7.0
1991	0.9	– 4.9	273.5	284.0	– 3.4	– 9.0	0.7

LONG TERM PERFORMANCE

Year End 31st Dec	Annualised % p.a. (Time Weighted)		IRR (Money Weighted)		Annualised % p.a. (Real)
	Fund	IPD	Fund	IPD	Fund
1980–91	9.6	10.0	10.1	11.6	3.2
1981–91	9.0	9.4	9.6	10.9	3.2
1982–91	9.2	9.7	9.8	11.2	3.3
1983–91	9.7	9.9	10.4	11.4	3.7
1984–91	9.8	10.0	10.6	11.5	3.7
1985–91	10.2	10.1	11.4	11.4	4.1
1986–91	10.6	9.9	12.0	11.2	3.9
1987–91	8.7	6.0	9.7	7.2	1.5
1988–91	2.5	– 0.5	2.7	0.4	– 4.3
1989–91	– 2.9	– 7.3	– 3.1	– 6.7	– 9.2

Note: 1 – Weighted average of all properties in IPD contributing funds.

2 – The IPD Index results shown here differ from the figures shown in the main part of the IPD Annual Review and the Property Investors Digest where properties which do not reflect normal market conditions are omitted.

EXTERNAL BENCHMARKS

Total return

Year End 31st Dec	Fund	IPD	Equities[1]	Gilts[2]	FTA Property[3] Shares
1981	15.2	15.3	14.3	2.4	5.0
1982	7.7	7.5	30.7	52.6	– 4.5
1983	5.4	7.4	28.4	16.2	33.7
1984	8.8	9.6	29.8	10.4	22.6
1985	7.1	9.2	19.8	12.6	8.2
1986	8.6	11.4	25.9	12.5	24.1
1987	18.5	26.6	7.1	16.4	23.4
1988	29.5	28.2	10.4	8.2	27.2
1989	14.3	14.7	36.0	7.4	5.7
1990	– 6.6	– 9.6	– 9.8	7.9	– 16.7
1991	0.9	– 4.9	20.0	18.2	– 11.9

Annualised

	Fund	IPD	Equities	Gilts	FTA Property
1980–91	9.6	10.0	18.6	14.4	9.4
1981–91	9.0	9.4	19.1	15.6	9.9
1986–91	10.6	9.9	11.7	11.5	4.0
1988–91	2.5	– 0.5	13.8	11.0	– 8.1

Income Yield Levels

Year End 31st Dec	IPD Net Income Yld	Equity Yield	Gilts Yield	RPI
1981	5.5	6.2	13.8	12.0
1982	5.8	6.3	15.5	5.4
1983	6.2	5.4	11.7	5.3
1984	6.5	5.1	11.3	4.6
1985	6.7	4.6	11.0	5.7
1986	6.7	4.4	10.8	3.7
1987	6.2	4.2	10.7	3.7
1988	.5.6	4.8	10.0	6.8
1989	5.8	5.5	10.2	7.7
1990	7.6	4.7	10.4	9.3
1991	9.0	5.8	10.8	4.5

1 – WM UK Equity weighted average Pension Fund Return.
2 – WM UK Bonds weighted average Pension Fund Return.
3 – FTA Property Share Price Index & estimated Income Return.

TOTAL RETURN BY SECTOR

% p.a. – Standing Investments Only

Year End 31st Dec IPD	Retail	Office	Industrial	Other
1981	17.8	15.2	12.4	13.7
1982	9.9	6.6	5.5	12.0
1983	11.9	4.7	5.4	11.9
1984	12.9	6.5	6.5	18.7
1985	11.8	7.4	3.0	18.5
1986	11.3	11.0	7.6	21.5
1987	18.4	30.4	21.4	22.5
1988	24.2	30.1	37.0	36.5
1989	9.2	15.4	26.9	17.8
1990	– 9.3	– 9.5	– 4.8	– 1.5
1991	2.9	– 11.3	7.9	– 5.0

Annualised

	Retail	Office	Industrial	Other
1980–91	10.6	8.9	11.1	14.6
1981–91	10.0	8.3	11.0	14.7
1986–91	8.4	9.4	16.7	13.0
1988–91	0.6	– 2.5	9.2	3.3

FUND

	Retail	Office	Industrial	Other
1981	26.0	19.1	5.7	–
1982	10.7	8.7	8.2	–
1983	7.9	5.4	5.9	–
1984	8.3	3.9	– 1.4	–
1985	7.2	6.2	2.3	–
1986	9.9	7.2	9.6	–
1987	12.5	16.2	15.2	–
1988	22.2	28.8	37.1	–
1989	10.9	10.9	23.8	–
1990	– 12.3	– 7.2	3.2	–
1991	4.7	– 4.4	7.7	–

Annualised

	Retail	Office	Industrial	Other
1980–91	9.4	8.2	10.2	–
1981–91	7.9	7.2	10.6	–
1986–91	7.0	8.1	16.8	–
1988–91	0.6	– 0.5	11.2	–

FUND NO PROPS

	Retail	Office	Industrial	Other
1981	11	24	12	0
1986	23	34	20	0
1987	26	29	18	0
1988	27	26	18	0
1989	26	26	17	0
1990	25	28	17	0
1991	26	28	17	0

FUND COMPARISON OF SECTOR PERFORMANCE

Index of Total Return (log scale) – Standing Investments Only

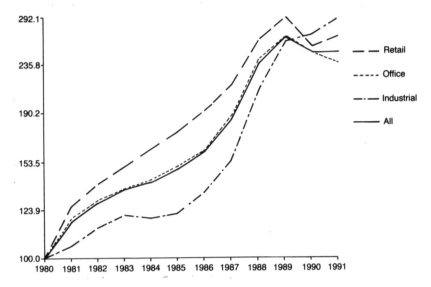

PERCENTAGE POINT DIFFERENCE IN TOTAL RETURN

Fund minus IPD

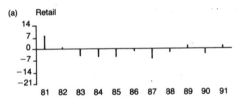

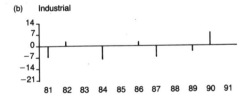

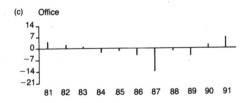

RETAIL SECTOR PERFORMANCE

Total return % p.a. – Standing Investments Only

Year End 31st Dec IPD	Capital Growth	ERV Growth	Equivalent Yield	Yield Shift	Income Return
1981	12.8	10.8	7.9	− 0.24	5.0
1982	4.9	5.6	7.9	0.04	4.9
1983	6.7	4.9	7.7	− 0.17	5.2
1984	7.6	7.0	7.5	− 0.04	5.3
1985	6.4	10.4	7.6	0.16	5.4
1986	5.8	11.7	7.7	0.31	5.4
1987	12.9	14.1	7.4	− 0.13	5.5
1988	18.9	21.1	7.2	− 0.06	5.3
1989	4.2	14.3	7.8	0.60	5.0
1990	− 14.7	5.3	9.4	1.62	5.4
1991	− 4.0	− 1.5	9.7	0.27	6.9

Annualised

1980–91[1]	5.2	9.3			
1981–91	4.5	9.1			
1986–91	2.7	10.4			
1988–91	− 5.2	5.8			

FUND

1981	19.6	32.5	5.1	− 1.69	6.4
1982	5.2	9.3	5.6	0.16	5.4
1983	2.7	3.5	5.6	− 0.02	5.2
1984	3.0	10.7	5.4	0.02	5.3
1985	2.0	1.7	5.4	0.04	5.2·
1986	3.8	22.5	6.3	0.67	6.0
1987	6.9	9.6	5.8	0.11	5.6
1988	16.8	16.3	5.8	− 0.12	5.4
1989	5.8	13.5	6.3	0.47	5.0
1990	− 17.5	2.2	7.8	1.51	5.2
1991	− 2.5	− 0.3	8.2	0.13	7.2

Annualised

1980–91[1]	3.7	10.7			
1981–91	2.3	8.7			
1986–91	1.2	8.1			
1988–91	− 5.2	5.0			

FUND NO PROPS

1981	11	2	2	2	11
1986	23	18	18	18	23
1987	26	24	24	24	26
1988	27	26	26	26	27
1989	26	26	26	26	26
1990	25	25	25	25	25
1991	26	26	26	26	26

Note: 1 – Annualised to base year of fund ERV series.

PERFORMANCE BY RETAIL PROPERTY TYPE

Total return % p.a. – Standing Investments Only

Year End 31st Dec IPD	Standard Shops	Shopping Centres	Retail Warehouses	Other
1981	19.6	14.6	15.2	18.5
1982	10.0	9.6	12.1	9.6
1983	12.1	12.2	8.2	11.2
1984	14.2	11.3	14.3	11.1
1985	12.9	10.7	11.2	9.6
1986	11.4	11.0	13.5	11.0
1987	20.9	15.9	15.3	15.4
1988	26.2	21.7	26.0	21.2
1989	8.7	9.5	10.7	8.9
1990	– 9.3	– 8.8	– 14.1	– 7.0
1991	1.8	1.3	13.0	7.2
Annualised				
1987–91[1]	6.1	5.3	7.9	7.1
1981–91	10.5	9.2	10.6	9.6
1986–91	8.9	7.4	9.3	8.7
1988–91	0.1	0.4	2.4	2.8
FUND				
1981	26.8	23.4	16.8	– 30.5
1982	9.6	18.7	13.9	107.4
1983	9.1	– 8.0	7.7	10.1
1984	7.4	28.8	12.8	5.1
1985	7.1	22.3	2.1	5.3
1986	7.5	16.8	11.3	16.2
1987	12.3	42.9	–	8.6
1988	23.0	–	4.7	25.1
1989	11.2	–	2.0	11.7
1990	– 11.0	–	– 8.2	– 25.2
1991	3.4	–	8.3	10.3
Annualised				
1987–91[1]	5.9	–	1.5	3.6
1981–91	7.7	–	–	13.9
1986–91	7.2	–	–	4.6
1988–91	0.8	–	0.5	– 2.7
FUND NO PROPS				
1981	8	1	1	1
1986	18	1	1	3
1987	22	1	0	3
1988	24	0	1	2
1989	23	0	1	2
1990	22	0	1	2
1991	22	0	2	2

Note: 1 – Annualised to the base year of the most recent fund series.

Summary of IPD performance analysis methods

The Purpose of IPD Portfolio Analysis

To produce as accurate a measure as possible of the return earned by the fund during the year on its directly owned property assets in order to provide the best estimate of the quality of the underlying assets and the skill of the management team in timing transactions.

The Accruals Principle

Since the timing of actual payments on a property portfolio is not always closely synchronised with changes in the properties which are reflected in the valuations, it is appropriate in general to adopt an accruals principle, which matches costs with benefits when recording cash flows. As far as possible, IPD will treat costs as accrued where a liability has been *incurred* although the money has not actually been paid or received. The prudence principle of accounting, which requires all *future* liabilities to be accrued will *not* be adopted.
Thus,

- *Transaction Costs*
  Transaction costs will be back-dated to the legal completion date of the purchase or sale at which the liability to pay was incurred. Where these have not been paid at the year end, estimated costs will be requested.
  Where agent of legal work on transactions is carried out in-house and not charged, standard scale fees for the work will be entered.
- *Retentions*
  If a fund has incurred a liability to pay a retention on purchase costs in the following year these will *not* be accrued as the liability to pay does not yet exist. These costs will be put through the accounts as and when they are paid.
- *Rents Receivable*
  Rents will be recorded on a receivable basis (as previously) i.e. when a rent review has occurred during the year the rent will be recorded as x months at the old rent and $(12 - x)$ months at the new rent.
- *Development Costs*
  Where funds accrue development costs in their book costs on items which have been built but not paid, and value these accruals, the book costs will be used as the basis of calculating the annual expenditure.
  If funds do not accrue these development costs in their own accounts the costs will be carried over into the following year. Where the fund's development valuations are based on book costs this will match valuation and expenditure. If, however, the development valuation has included the

accrual (although it was not included in the accounts) the performance will be artificially brought forward by one year. IPD will be forced to accept this anomaly as it would be impractical to change clients' accounting procedure. IPD would, however, encourage funds wherever possible to attempt to match their valuation and accounting principles.

- *Outstanding Rent Review*
Where a rent review is outstanding the estimated rental value at the date of the review will be used in place of the new rent in calculating rent receivable. The new rent will be introduced when settled but not backdated to avoid changing historic results.

- *Lease Expiries*
If a lease has expired the old rent will be used until a new lease is in place.

- *Rolled Up Interest*
Where funds have entered into funding agreements with developers and the interest owed by the developer has been rolled up, it will not be included in the fund's expenditure recorded by IPD. IPD would not expect rolled up interest to be included in costs and require to be notified if that is the case.

- *Actual or Notional Interest on Direct Developments*
In recording the performance of directly owned property, IPD excludes from performance measurement any element of gearing. If either actual or notional interest is included in the book costs this should be notified to IPD and will be deducted. If the development valuation is based on book costs which include either of these items this must be discussed with IPD.

- *Service Charge & Other Revenue Receipts*
IPD will record service charges (including management costs where they are recoverable) and all other revenue receipts such as wayleaves or developer guarantees.

- *Ground Rents*
Ground rents will be recorded on a payable basis i.e. the amount which was due to the superior landlord during the year.

- *Management Costs*
IPD will record only "those costs incurred in maintaining and improving the rental income on a property and that are charged against revenue". This includes:
– Basic rent collection fees
– Rent review and lease renewal fees
– Other building specific fees charged to revenue
Management costs will *exclude* portfolio management fees and valuation costs.
    Where a fund conducts management work in-house and no internal fee is charged for any of the above items, IPD will apply standard scale fees (an average of agent fees supplied by IPD Sponsor firms).

Where management fees are recoverable on service charge properties they will be excluded by the recording of service charges as a revenue receipt (see below).

- *Other Revenue Expenditure*

All other expenditure (excluding ground rents and all management costs) which is charged against revenue will be deducted.

These principles have been agreed by consensus amongst our major clients and will, we feel, provide the most useful performance measures for the purpose of property portfolio management. It should, however, be noted that the **use of the accruals principle does not result in a perfectly accurate return on capital employed and will not exactly match internal results calculated on a cash paid basis. Differences may also arise from the investments included in the analysis.** (See below).

Properties to be Included

IPD record only directly owned properties in the UK and Eire (the latter converted to sterling).

Agricultural properties and forests are also included but their results are shown separately and the effect of their inclusion in the portfolio will be shown in the Agricultural section of the reports.

Equity share properties are not included in standard reports but their inclusion can be analysed as an optional extra by agreement with IPD.

Property shares, holdings of Property Unit Trusts and cash are not included.

Owner Occupied Properties: Where these are included in investment portfolios and are valued at Open Market Value, they will be included. If an actual lease is in place to the occupying company and a market rent is charged, this figure will be used. If a discounted or hypothetical rent is used in-house, IPD will substitute a rent equal to the average of the ERV two and three years previously. Where insufficient historic evidence exists, IPD will discount the current ERV by $2\frac{1}{2}$ years on the basis of the appropriate IPD Monthly Index sector results.

Index

OTHER TITLES FROM E & FN SPON

Management, Quality and Economics in Building
P. Brandon and A. Bezelga

Property Investment and the Capital Markets
G. Brown

Risk, Uncertainity and Decision-making in Property Development
P. Byrne and D. Cadman

Property Development
3rd Edition
D. Cadman and L. Austin-Crowe
Edited by R. Topping and M. Avis

Microcomputers in Property
A surveyor's guide to Lotus 1–2–3 and dBASE IV
T. Dixon, O. Bevan and S. Hargitay

Caring for our Built Heritage
A survey of conservation schemes carried out by County Councils in England and Wales
T. Haskell

Rebuilding the City
Property-led urban regeneration
P. Healey, D. Usher, S. Davoudi, S. Tavsanoglu and M. O'Toole

Property Investment Theory
A. MacLeary and N. Nanthakumaran

National Taxation for Property Management and Valuation
A. MacLeary

Industrial and Business Space Development
Implementation and urban renewal
S. Morley, C. Marsh, A. McIntosh and H. Martinos

Property Valuation
The five methods
D. Scarrett

Investment, Procurement and Performance in Construction
The First National RICS Conference
P. Venmore-Rowland, P. Brandon and T. Mole

Industrial Property Markets in Western Europe
B. Wood and R. Williams

Journals

Journal of Property Research
(Formerly Land Development Studies)
Editors: B. MacGregor (UK), D. Hartzell and M. Miles (USA)

For more information about these and other titles published by us, please contact: The Promotion Department, E & FN Spon, 2–6 Boundary Row, London, SE1 8HN